And You Know You Should Be Glad

BOOKS BY BOB GREENE

And You Know You Should Be Glad

Fraternity: A Journey in Search of Five Presidents

Once Upon a Town: The Miracle of the North Platte Canteen

Duty: A Father, His Son, and the Man Who Won the War

Chevrolet Summers, Dairy Queen Nights

The 50-Year Dash

Rebound: The Odyssey of Michael Jordan

All Summer Long

Hang Time

He Was a Midwestern Boy on His Own

Homecoming: When the Soldiers Returned from Vietnam

Be True to Your School

Cheeseburgers

Good Morning, Merry Sunshine

American Beat

Bagtime (with Paul Galloway)

Johnny Deadline, Reporter

Billion Dollar Baby

Running: A Nixon-McGovern Campaign Journal

We Didn't Have None of Them Fat Funky Angels on the Wall of Heartbreak Hotel, and Other Reports from America

BOOKS BY BOB GREENE AND D. G. FULFORD

To Our Children's Children Journal of Family Memories

Notes on the Kitchen Table: Families Offer Messages of Hope for Generations to Come

To Our Children's Children: Preserving Family Histories for Generations to Come

BOB GREENE

And You Know You Should Be Glad

A True Story of Lifelong Friendship

WILLIAM MORROW
An Imprint of HarperCollins*Publishers*

HarperCollins books may be purchased for educational, business, or sales promotional use. For information please write: Special Markets Department, HarperCollins Publishers, 10 East 53rd Street, New York, NY 10022.

FIRST EDITION

Designed by Jeffrey Pennington

Printed on acid-free paper

Library of Congress Cataloging-in-Publication Data

Greene, Bob.
And you know you should be glad: a true story of lifelong friendship/ Bob Greene.—1st ed.
p. cm.
ISBN-13: 978-0-06-088193-1
ISBN-10: 0-06-088193-3
1. Greene, Bob. 2. Greene, Bob—Friends and associates. 3. Greene, Bob—Homes and haunts—Ohio—Bexley. 4. Authors, American—Homes and haunts—Ohio—Bexley. 5. Authors, American—20th century—Biography. 6. Bexley (Ohio)—Biography. I. Title.

PS3557.R37963Z464 2006
977.1'56—dc22
[B]

2005054125

---

06 07 08 09 10 WBC/RRD 10 9 8 7 6 5 4 3 2 1

For you, who are reading this . . .

and for your best friend

And You Know You Should Be Glad

One

WE WALKED SLOWLY TO AUDIE MURPHY Hill.

It's at the corner of Ardmore and Elm—the north edge of the small front lawn at 228 South Ardmore. He—Jack—used to live in that house, when we first became best friends. We were five then; we were fifty-seven now, standing next to the lawn, next to Audie Murphy Hill.

"It seemed so steep," I said to him.

"Well, we were little," he said.

This was toward the end—there would not be many more of these walks for us, the months leading up to today had taken their toll—but every time I came back home to see him, we made the walk. He wanted to.

The slope hardly rises at all—it's not really a hill, at least in the eyes of grown men. But in those years when he and I first knew each other—the years just after World

War II, the years during which the fathers of the families in the town had come home from Europe and the Pacific, had bought houses on streets like this one, had started to settle back into life during peacetime—it had felt to us like something out of Italy or North Africa. We would charge up that slope—up that placid piece of grass on that safe Ohio street in a town where only 13,000 people lived—and, sticks in hand, sticks standing in for rifles, we would pretend that we were Audie Murphy. The most decorated combat soldier of the Second World War.

"Maybe the new owners of the house leveled off the lawn," I said to him now.

"No," he said. "This is how it was. It just felt steeper."

We were still on the sidewalk. I was trying to see what was in his eyes, without him knowing I was looking. Fat chance. He always noticed everything.

"Your dad used to watch us sometimes," my oldest friend—no longer a boy, no longer sure of anything—said. He was getting tired. I had told his wife that we wouldn't be long. Their house was less than a mile from his parents' old house—less than a mile from Audie Murphy Hill.

"I know," I said. "My dad would be picking me up in his car, to take me home for dinner."

Those men home from their war—what must they have thought? It hadn't even been ten years for them, back then—ten years earlier they had been fighting in Europe, fighting on the islands of the Pacific, and then they were here, leaning against their Fords and Buicks, waiting while their sons finished playing soldier in the dying sun.

"They were much younger than we are now," I said to Jack.

"They were in their thirties," he said.

I thought I should ask him, so I did:

"You feel like climbing up the hill?"

It wasn't a hill at all. But it was too steep. Now, near the end, just as at the beginning of our lives, at the beginning of our friendship, it was too daunting for him, at least on this day.

"Let's go back," said my oldest friend.

We started to walk—slowly, because he was unsteady—toward his waiting wife, toward home.

Two

We all, if we're lucky, have someone in our lives like Jack—our first friends, our oldest friends.

If we're especially fortunate, they remain close to us no matter where the world leads us. We don't have to live in the same cities; we don't have to see each other on a daily basis. Friendships—especially the oldest friendships—don't require that.

No one knows us better. No one in our adult lives saw us the way we first were, before the inevitable defenses against a thorny world went up, before the layers of protective walls around us were constructed. We didn't invite the arrival of those defenses; we didn't willingly participate in the building of those walls. They come, eventually, with life—included in the package.

We all have someone who was there before all of that.

If we're lucky, the someone is with us for a very long time.

I KNOW THE EXACT MOMENT I MET HIM.

We were in kindergarten, at the Cassingham Elementary School in Bexley, Ohio, about a block and a half from his house.

Those first weeks in kindergarten are confusing—thrilling, and a little overwhelming, but mostly confusing. For the first time, you're on your own. You've been dropped off at school the first day, and suddenly you find your new world unfolding in a building other than your parents' house. You're the youngest in the school—you're five, and on the stairways sometimes you see people as old as twelve, laughing and talking loudly and clearly familiar with each other in a way that you're familiar with no one. In an unfamiliar place like that you can feel pretty small. Especially because you are.

The classroom, on the south end of the first floor of the building, was filled with keyed-up noise most mornings. Everything was new, every day—the layout, the snack routine, recess, the bells out in the hallways . . . nothing seemed remotely like anything you had experienced before.

Miss Barbara was the teacher. She was just a few years out of Ohio State University—Barbara Drugan was her name, and although to us she seemed as old as someone in a history book, she was still in her early twenties.

We had all, on the first day, been told each other's names, but those dozens of names were hard to remember; in the initial weeks, Miss Barbara asked us to call out our own names every day as she took attendance. I think she probably was doing it for a reason—she already knew who was there and who wasn't, she was having us do the roll call out loud so we would gradually become recognizable to each other. Put names to faces, for the first time in our lives.

So we sort of knew—we didn't know every name in the classroom on Cassingham Road, but we were getting there, day by day.

One afternoon we were all sitting on the floor, in a semicircle around Miss Barbara, who was reading to us. I was sitting near the back of the group of children, and I thought I noticed something on my upper lip—it felt as if my nose was dripping.

I didn't have a Kleenex or a handkerchief, so I lifted my hand to my face to wipe away whatever was there. It didn't seem to work; I swiped my hand against the top of my mouth, but my upper lip felt just as wet afterward.

I looked at my hand. It was red. I was bleeding.

I'd had nosebleeds before, but at home my mother was always close by to take me into the bathroom and help me make the bleeding stop. She would hold a piece of tissue to my nose, and hold my head back, and sometimes, if she needed to, she would get a piece of ice.

Now I was sitting on the floor at the back of the semicircle of other five-year-olds, and I put my hand to my face again, and it came away covered in blood.

I was embarrassed. I didn't know what to do. When you're that age, the last thing you want is to be singled out in public because something is wrong with you. You just want to fit in with everyone else, to be lost in the crowd.

I put the other hand up to my face, trying to do this as quietly as I could. I only wanted the bleeding to stop.

I tilted my head down toward the linoleum floor, hoping no one would see what was going on. I lifted the bottom of my T-shirt to my face, pressed it against my nose, thinking that if I did that, the pressure would stop the blood.

It didn't. Now the shirt had blood on it, and the more I attempted to stop the bleeding, the more bloody the shirt became. I was beginning to feel a five-year-old's panic; the blood was coming out faster now, it was all over me and all over my clothes, and I had no idea what to do. I felt as if this was somehow my fault. I bent over even further, trying to disappear into the floor, doing everything I could to let no one see me.

Then, a few feet away from me, someone stood up.

I heard his voice before I saw him. I had been staring straight down, scared and ashamed.

"Miss Barbara?" I heard the voice say.

She stopped reading aloud.

"Bob's hurt," the voice said.

His name was Jack Roth. We didn't know each other, but he had been listening during the daily roll call, and had learned my name.

He stood there looking at Miss Barbara, and she looked at him, and then over at me—blood all over me—and within a minute I was down in the school nurse's office, getting help, getting cleaned up, being told not to worry, everything was going to be fine.

"Bob's hurt." He hadn't hesitated even a second. Like the rest of us, the last thing in the world he wanted was to disrupt this new kindergarten world we were all just getting used to; like the rest of us, he was feeling his way every day, learning the customs, figuring out the rules. Standing up and interrupting the teacher was not something that came easily, especially that early along the path on which we were just starting out. When the teacher was speaking, you didn't.

But there he was. Standing straight up, for someone he didn't yet know.

AT FIFTY-SEVEN, ALL THESE YEARS LATER, we walked from Audie Murphy Hill to the house where he now lived, the house where his wife was waiting.

For most of our young lives, in the hours after school, we would go either to his house or to mine. In the early years, our mothers would almost always be home; in the 1950s, at least in this pocket of central Ohio, mothers who were stay-at-home housewives were the rule, working mothers were the exception.

My mother would have egg salad sandwiches ready for us; his mother would have Toll House cookies. I had never heard those words in combination before—"Toll House," as a brand and as a concept, was a stranger to the house where I grew up, the words might as well have been a foreign phrase until I met Jack—but the afternoon aroma of those cookies baking in the oven in his mother's kitchen was the sensory signature of that house on Ardmore. Out on the gentle slope at the edge of his front lawn we might have fooled ourselves into thinking we were American combat soldiers—but we were combat soldiers who, when the battle was over, returned not to foxholes or mud-splattered tents, but to chocolate-chip cookies just a few steps inside the back door.

Jack's mother died when he was fifteen. We who were his closest friends knew she had been sick, but we'd had no hint she had been that kind of sick. There was a little paid death notice in the paper saying that Mildred Roth of 228 South Ardmore Road in Bexley had passed away, and from that day forward when he went home in the afternoons it was to an empty house. If he ever felt cheated by that—if he ever felt lonely, or let down, or hollow inside, if he ever walked in that back door and turned on the lights and knew before he consciously knew it that the comforting scent of those cookies was no longer present, that it wasn't coming back—if he ever went to his room and closed his eyes and cried, he didn't tell us about it.

"Bob's hurt," he said that day when we were five.

But when he was hurt, he kept it inside.

IF YOU ARE NOT PARTICULARLY FAMOUS in the eyes of the world—even in the local eyes of your local world—then the chances are dim that your passing will be noticed.

If your accomplishments are quiet accomplishments, cherished only by those who love you, likely you will go in silence.

The friends who mean everything to us—the friends without whom our lives would be empty—are our most enduring models of grace and good fortune. When we lose them—and we all do; we all will—we realize, then and forever, that our own lives have been filled to overbrimming with the grand, invisible gifts they have given us. We know that our time on earth would have had paltry meaning had not, one fine day, our lives connected for the first time with the lives of the people who would turn out to be our most treasured friends.

Jack didn't get a news obituary in the local paper—just

a paid death notice in agate type. His life, like that of his mother, did not attract the attention of a reporter working for a daily newspaper.

You didn't know him. But there is someone in your life very much like him.

Someone . . . and something.

This is the biography of a friendship.

Someone could write it about the dearest friendship in your life. If you're lucky—and you probably are—that kind of friendship is there, or it was, once upon a wondrous time.

Three

There were so many telephone calls over all the years—calls made on heavy black Bakelite rotary-dial phones when we were children; calls made on sleek and too-light Trimline touch-tone phones when those became the style in American homes; calls made on multiline business-office phones with flashing hold buttons when we had our first jobs; calls made on cell phones when those became a ubiquitous, and perhaps ultimately regrettable, part of the national landscape.

No matter where we were, whether it had been weeks or merely hours since we had last talked, there were two days of every year when we would always, without fail, ring each other's phones. Both days were in March.

On March 10, every March 10, he would call me to wish me a happy birthday. On March 25, every March 25, I would call him to do the same.

On the March 10 that was my fifty-seventh birthday, he called with a chuckle in his voice. "You think fifty-seven is bad, wait until we're seventy," he said. Across the miles, I could almost see the smile.

Neither of us knew.

Two weeks later, he turned fifty-seven. It would be his last birthday.

I WAS WALKING ON A BEACH IN FLORIDA late that March; my own life had hit some rough waters, and I was trying to sort things out. I had come down to the Gulf of Mexico for a few weeks to walk in the sun and spend some time thinking. That's where Jack had found me with his March 10 birthday call; that's where I had been when he had joked about what it would be like when we turned seventy.

Several weeks had passed since then. I had walked to the end of Longboat Key, touched the bottom of my shoe to the same boulder I touched every day before turning around to begin the walk back, and was about ten minutes into the return leg when I stopped to look out at the Gulf.

The sun was blazing high in the sky, there were only wisps of clouds, the water was so clear you could see the sand beneath it. I had a cell phone in my pocket; it had not rung.

What kind of idiot carries a cell phone with him on a walk down the beach? Life must have gone on quite smoothly before we were reachable every second, before we had the option of constantly checking to see if anyone was looking for us. But check we do, mainly because we can.

I had resisted checking my home telephone up north in Chicago—why interrupt a sun-drenched and becalmed morning like this to search for reminders of a grayer world? But with my cell phone on the quiet beach, with the soft and soothing sound of the Gulf's waves in my other ear, I called the proper number in Chicago, punched in the requisite code, and learned that I had one new message.

"Nice answering machine," the voice said. "This is Chuck."

He was referring to the beep. My wife's voice had always been the voice on our home voice mail. After she died, and I realized that her voice was still the one greeting callers, I took it off the machine. I didn't replace the greeting. I just let the beep, and no voice at all, answer the phone. I'd left it that way for quite a while.

"Nice answering machine," Chuck said.

The reason I know his exact words, and the ones that followed, is that I have saved that call on my voice mail. Some calls change things forever.

THERE WERE FIVE OF US WHO WERE BEST friends. There was Jack. There was me. There was Chuck Shenk, and Danny Dick, and Allen Schulman. In high school, back in Ohio, we had called ourselves ABCDJ. Allen, Bob, Chuck, Dan, Jack.

I could probably count on the fingers of my hands—maybe the fingers of one hand—the number of serious conversations that Chuck and I had ever had. With us, everything was always laughter. The way I usually explained it to people was that with Chuck, when we were boys, it came down to this: You could tell him that the world was going to end at midnight, and he'd probably just toss his head to the side and then say, "Let's go to the Toddle House for cheeseburgers and banana cream pie." We'd grown older over the decades since then, but that hadn't changed.

I know the varying tones of his voice—even over a phone line, even on a recorded message coming through static into a cell phone on a Florida key removed from the mainland, I know the tones of Chuck Shenk's voice. And in those first few words—"Nice answering machine. This is Chuck"—I could tell that something was off-kilter.

"Give me a call," the voice continued. "It's about Jack. He's a little ill, and I wanted to explain it to you. So give me a call. Bye."

This wasn't the way Chuck ever spoke to me. *He's a little ill, and I wanted to explain it to you.* Chuck had never couched a sentence in words like that in our lives.

With gulls overhead and the water sparkling out to the horizon, I made the call.

JACK'S WIFE HAD YELLED UPSTAIRS TO say dinner was ready.

That's what Chuck told me when I reached him.

Jack apparently hadn't been feeling well for a few weeks; he thought he might have the flu. He came home from work early and went up to the bedroom to take a nap. He'd told Janice, his wife, to let him know when it was time to come down for dinner.

She had called up the stairs, Chuck said, and Jack hadn't answered.

She had called a few more times, and then had walked up the stairs to see if he was asleep.

She found him on the floor, unconscious.

THE GREATEST COMPLIMENT ANYONE EVER paid our friendship, I think, was in the second grade, when Miss Hipscher moved us apart.

She'd had enough of our constant talking. Jack and I had chosen desks next to each other—they were those desks where the flat-surface writing part is attached to the chair part, you had to slip into them through the opening on the side—and after a few weeks of watching us talk to each other and joke with each other and not pay attention to what she was trying to teach us, Miss Hipscher called us aside one morning and said she had made a decision.

She said she could tell that we were good friends—we were such good friends that she was going to move us to desks in different parts of her classroom. She wasn't trying to be unkind about it; she made it seem perfectly logical. She told us that we were never going to learn anything if we sat next to each other and talked all day. It was better for us, she said, if she put some distance between us.

What a great thing for a person to notice about a friendship. What better testament than that: You two are such good friends that I have to move you apart.

On the Florida beach I listened as Chuck told me the rest of the story. How the Bexley police and the emergency squad had come to Jack's house, had carried him down the stairs. Something about the hallway turn from his bedroom being too narrow for a regular stretcher, and them having to lift him onto a folded sheet to carry him to the waiting ambulance.

We'd been moved away from each other in that classroom so many years before. For our own good, the teacher had said. I looked at the Gulf and I had never in my life felt so far apart from my oldest friend.

Four

OF COURSE IT WOULD BE ON A BUS.
That's when the moment would come.
Only he and I would understand.

THE FIRST TIMES WE FELT LIKE WE WERE people in the wider world—the first times we felt untethered, independent—were the times when, nickels in hand, we would stroll off toward the bus stop, to wait for the Bexley bus.

It makes me grin, now. The Bexley bus didn't go anywhere. It never left the town. It just—for a nickel, in the 1950s—made this continuous, languorous, irregularly shaped loop through our little town that had no manufacturing plants, no industries, no tourist attractions. The bus never left the Bexley limits.

Bexley, small as it was and is, is surrounded on all sides by the City of Columbus. Head in any direction and you'll end up in the bigger town. But the Bexley bus never headed far enough in any of those directions. Just when it seemed that the bus might lurch a foot or two over the line and step out of town, the driver would steer a sudden right or left, and the loop would continue, as always.

That's why our parents, when we were seven or eight, permitted Jack and me to ride the Bexley bus alone. We weren't going anywhere. The Bexley bus had as much in common with a ride at Disney World—not that there was any such thing as Disney World, or even Disneyland, back then—as it did with conventional municipal public transit. It looked like a bus and felt like a bus and smelled like a bus—but it didn't go anywhere.

Which made it perfect for two boys wanting to be out of the house and on their own for the first time. Jack and I would walk to the bus stop on East Main Street, wait for the bus with the "Bexley" sign in the tall front window, listen for the pneumatic sound of the rubber-edged door creaking open . . . and then we were on.

Up those narrow steps, drop the nickel into the slot—the clatter of our nickels clanking into the glass-sided collection bin next to the driver's seat was, to us, the exhilarating, almost breathtaking, sound of newfound freedom. A quick and high-pitched hello to the uniformed driver (what must he have thought every morning as he put on that freshly pressed gray uniform and cap, knowing that all day long he would have only a handful of riders, all of them going nowhere?), and we were on our way.

Many summer afternoons we were the only customers. We would pick a seat halfway back. Even when we were the sole riders in the middle of the day, even when we each could have had half the coach for ourselves, we sat together, side by side, as if this was a crowded-to-capacity commuter bus in a huge metropolis, as if seating was difficult to find and valuable beyond measure.

So there we'd be, two little boys on a worn-leather bench seat, riding up Roosevelt Avenue, riding up Cassady, knowing every stop. We'd talk . . . that whole ride we would look out the window (as if we were going to see something we hadn't seen before), and talk with each other about what was out there, feeling silently thrilled that we were doing this. The Shell and Sohio gas stations would pass in and out of our line of sight, and Soskins Drugstore, and the Eskimo Queen ice cream stand and the high school with its football stadium out back and the Kroger grocery store and the stone front of the Bexley Public Library . . .

The driver would call out the stops, just for us. I wish I had a movie of it now. We felt we were out in the world, yet the world, by design, was so protected. Nothing was unknown. The Drexel movie theater would pass in and out of our window, and the Feed Bag lunch counter, and Seckels 5 & 10 Cent Store, and the houses—all those Bexley houses, with all those trees in front—and when the meandering loop was completed, when we were right back where we had started, the driver would pretend not to notice we still were there.

By pretending not to notice, he didn't have to ask us for another nickel—we could continue to ride the endless loop. He seemed just as glad to have the company—just as glad to have someone, as opposed to no one, onboard. Even if it was only us. We rode that bus together, on our own, or something like it.

CHUCK HAD A BUSINESS FRIEND WHOSE company owned a bus—one of those semi-luxurious tour-type buses that take people to parties or sports events.

On the first weekend after the doctors in Columbus were able to bring Jack back to consciousness—after they had stabilized him (such a cold and ultimately misleading word) and had run him through the tests and determined that he was full of cancer, including in his brain—Chuck had implored his business friend to lend him the bus and its driver.

Jack's family had gathered. Jack—he was awake now—wanted to go see Lance Armstrong's doctor. He had read the book—he had been inspired by Armstrong's victorious battle against cancer. The doctor—Lawrence Einhorn of the Indiana University Medical Center in Indianapolis—had agreed to give Jack an appointment at the beginning of a weekend, even though Dr. Einhorn was supposed to be going out of town.

Jack was in no shape for a car ride over to Indiana—he was coming straight out of the hospital in downtown Columbus—and an ambulance, everyone agreed, was best avoided. So Chuck—more serious and determined than I have ever known him—remembered his business friend with the bus.

The Bexley bus had never strayed beyond the protective boundaries of our quiet town. Now Jack was headed out of state, final destination to be determined.

I HAD HEARD SCREAMING.

This was just before the Indiana bus trip; this was when Jack was still in Columbus, in those first few days after he collapsed. I couldn't figure it out. I knew he wasn't the one screaming; his voice, from his hospital bed, was coming through the phone at the same time the screams sounded. He had been telling me about the preliminary diagnosis: cancer in his brain, cancer in his lungs, cancer in his liver. This, out of nowhere. To a guy who never smoked, to a guy who watched his diet like some sort of punctilious nutritional accountant, to a guy who tried to go for a long run every day of his life.

"What is that?" I asked him. I was still in Florida, waiting to find out what was going on.

"The noise?" he said. He was groggy from the sedatives they'd given him before the biopsies.

"It sounds like someone got shot or something," I said.

"Someone did," he said. "That's my roommate."

They'd put him in a room with a man who'd taken a bullet to the head. No single rooms at the hospital in Columbus were available; there was just a curtain between Jack and the gunshot victim. Jack had to hear that he had cancer, and absorb what that meant, while sharing a cramped room with a guy who was yelling from the pain of a shot to the skull.

"Can't they get you something better?" I asked. "Something with a little privacy?"

"This isn't so bad," Jack said. "Most of the time I just tune him out."

Not angry even at a moment like this. Getting the diagnosis from the doctors, cooped up in a hospital room with a man shrieking from a gunshot in his head . . . Jack was Jack, even with this.

But then, that should have been no great revelation. The first of us to have no mother, then the first of us to have no mother and no father—first to be parentless—and, all these years later, I can tell you exactly how many times we heard Jack complain about that:

Zero.

Never. Not once.

So, in the hospital room with the soundtrack of excruciating pain in the background, he told me to wait before I came to Ohio. He was going to Indiana, he said. His voice, sedated as it was, was laced with hope. Lance Armstrong's doctor had agreed to see him. And Chuck had come up with a bus.

THE NEXT TIME I HEARD HIS VOICE, I barely recognized it.

He'd always done such a good job of keeping whatever unhappiness there was in his life inside. I was all but unfamiliar with what he sounded like when he was scared and full of sorrow.

But there it was—once more on that most antihuman and unsatisfying form of one-way discourse, the voice-mail message taker.

It was on the day of the bus trip to Indiana.

I know what he said word for word, because—like that

first message from Chuck—I've saved it. Not that I can stand to listen to it; it's been there for almost a year now, and Jack's gone, and I made myself play it a minute or two ago so I would get it right as I type this. I never want to lose his voice, but I can hardly bear to hear it.

This is what he said that afternoon, after leaving Dr. Einhorn:

"Greene, it's me. Oh, man. Bad news today, buddy. Give me a call when you get a chance. I'll talk to you later."

The sound of that wounded voice. . . .

"LITTLE HELP?"

Two words we'd been hearing all our lives. As kids on the baseball diamond behind Cassingham Elementary, we'd say the words or we'd respond to them, depending on whether we'd let the ball slip away from us, or someone else had, on a different part of the playground.

"Little help?" Meaning: Give me a hand—toss the ball back to me. The ballfield shorthand of American boyhood.

We'd sense a ball skittering past our feet, and then, from out there on another patch of grass, we'd hear the call of someone's voice. "Little help?" We'd bend to retrieve the ball, look to see from where the voice was coming, and throw it back.

Later, when we were grown men walking together through Bexley, there would be times when we'd hear it once again. The words were like long-lost echoes, reminders of something we'd almost forgotten about. We'd be walking, in our twenties and thirties and forties, down

Powell or Stanwood or Remington, and the words would arrive from a front yard, like a faded postcard found and delivered after being stuck for decades in a crack in the post office sorting room. We'd be walking—men, by then—and the sound would reach us:

"Little help?"

We'd see a ball in the street, and one of us would kneel to pick it up. There'd be a kid in one of the yards, waiting.

When you're older, it's much more difficult to ask. When life's troubles get more serious—when your problems go beyond the fleeting inconvenience of an errant baseball—you tend not to say what you need, not out loud. You almost never can bring yourself to say to someone: "I have a problem, I need help." The instinct to do that has been trained out of you by life.

It used to be routine, when the problems were easily fixable.

"Little help?"

We could use some now. That's what I thought, as I got ready to return Jack's phone call. We could use a little help right now.

EVEN ON THE WORST DAYS—THE DAYS when you learn that nothing will ever be the same again—the mundane necessities of existence require tending. No one had eaten since early that morning. In Columbus, Jack, Chuck, their wives (Jack and Chuck married twins, twins from Cleveland—I know, what are the chances?), Jack's twenty-four-year-old daughter, who had come in from her job in New York, Chuck's older daughter . . . all of them had boarded the bus for Indiana, and now it was the middle of the afternoon and they were on their way back home, and they hadn't eaten. Not that anyone felt like it.

But they had to eat something, so the bus had stopped at a restaurant by the highway and everyone except Jack had gone inside to order carryout sandwiches to bring back. If you hadn't known, the whole thing might have looked from the outside like an excursion to Ohio Stadium for a Saturday football game.

I reached Jack while the rest of them were inside the sandwich place.

Dr. Einhorn, he said, had been lovely. Kind, and thorough, and willing to spend as much time as Jack needed. Jack had brought his X-rays from the tests in Columbus with him, and the results of all his lab work. Dr. Einhorn, Jack believed, was the man to go to when you were seeking miracles. Look at Lance Armstrong.

It was with sadness and realism that Dr. Einhorn told Jack and his family that if Jack would begin radiation and chemotherapy immediately—and he meant immediately, Monday—he might have a chance to live for a year or two. This was a level and a type of cancer that was not going to go away. The doctor wished he could say otherwise, but he couldn't.

"He's really a nice guy," Jack said. "You could tell that he knows exactly what he's talking about."

I didn't have any words. I could hear voices; the others were getting back onto the bus with food for the rest of the grim ride back.

And here is what Jack said to me to sum up what he had just been told. A month before, he had been working hard at his job every day, he had been laughing with his friends and family, he had been making plans for vacation trips, and running and working out to keep himself in shape, he had been a man on top of the world. It hadn't even been three weeks since he had laughingly said to me, on my birthday: "You think fifty-seven is bad, wait until we're seventy."

Now he had been told what he had been told, and here, in a clear and steady voice, is what he said to me:

"I got dealt a hard hand."

That was it, and the other voices became louder as the rest of them made their way up the aisle of the bus.

"This isn't a good time to talk," Jack said. "I'll be back at home tonight."

I said I'd be there soon. We hung up. I thought about the Bexley bus, aboard which there were never any surprises, never any unexpected turns.

Five

I WENT TO THE AIRPORT IN CHICAGO, picked up my boarding pass, stopped in at the newsstand. I took my seat on the plane, looked out the window during the flight, waited for my bag on the carousel inside Port Columbus. I caught the hotel van to the place where I'd made my reservation, checked into my room, tossed my suitcase onto the second bed, the one I wasn't going to use.

I did all of this in exactly the way I'd done it a thousand times before. I did it the same way I would do it on any other trip, because, while I still could, I suppose I wanted to fool myself into thinking that this was any other trip.

IN THE FIRST GRADE—AT LEAST I THINK it was first grade, it may have been later, because I remember it was in the schoolroom in which Jack and I made up new words to the song "Sixteen Tons" by Tennessee Ernie Ford, and "Sixteen Tons" may have come out after we were in first grade, so I may have the elementary school year wrong . . .

Whatever early year of school it was, the year that Jack and I rewrote "Sixteen Tons" to sing to the class about arithmetic or maps or something, there was an afternoon when the school nurse came to the door of the classroom.

She said something to the teacher. Then she and the teacher walked to the desk of a girl, a very nice girl with whom Jack and I were friends. They said something quietly to the girl, and walked her out of the classroom, and that was the last we saw of her for several weeks.

What they had told her—we found out later that day—was that her mother had died. It didn't really register with us, what that could mean. Death had not yet entered our lives, it would be years before Jack's own mother would die, so he and I sat there in class, and we looked at each other, and that day after school maybe we talked about it a lot and maybe we didn't. But something had changed.

From my hotel room on the outskirts of the Columbus airport I called a cab to take me into Bexley to see Jack, and maybe that day in the classroom had something to do with why I was not feeling all worked up, why I was feeling mostly numb. One minute Jack and I had been writing goofy new lyrics to "Sixteen Tons," and the next minute we found out, for the first time, that someone's world can change in the blink of an instant. They walked that girl right out of the classroom and into the rest of her devastated life. I guess I'd been waiting for that to happen again ever since.

Usually it didn't; sometimes it did. I waited by the front desk of the hotel and the cab pulled up, not a moment late, right on time.

I TOLD THE CABDRIVER THE EXACT DIRECTIONS I wanted him to take, street by street, because before I saw Jack I wanted to ride by his old house.

So we took James Road from the airport south to Broad Street, and then Broad Street into Bexley one block past Cassingham, and I asked the driver to take a left on Ardmore and follow it on down to Elm.

It was raining in Bexley that late afternoon, and the grass on Audie Murphy Hill was slick and dark, and I looked up to the second floor of the old house, to his bedroom window.

We had taped a cardboard box above his bedroom door; we had used Scotch tape to secure it against the wall, we had to use a lot of tape because the box was going to get a lot of use. We had cut the bottom out of the box, so that anything you dropped through the top would fall right through.

That was the point. We rolled up pairs of Jack's socks, and we played bedroom basketball. After school, day after day, that's what we would do. We would feint and lunge, we would try to fool each other with moves, we would put our backs to the basket and attempt hook shots—all of this while we had those rolled-up socks in our hands, the rolled-up socks serving as basketballs.

We'd sweat—it doesn't sound as if something you'd do in such a confined space would make you sweat all that much, but we worked at it, we loved those games. We'd keep score. The laughter in that room, the shouts of triumph or defeat during the games . . . it warmed our winter days. Things of great significance may have been going on in the world outside the walls of Jack's house—politics, crime, fodder for headlines—but our laughter and our scorecards, for hours at a time, were a complete and self-contained universe of their own. We needed nothing else.

And, from downstairs, the scent of those Toll House cookies, baking even as I would play defense and Jack, looking left, looking right, would jump and loft the balled-up socks toward the top of the cardboard box. We pretended that we were opponents; we pretended that we were on different sides.

From the cab I could see the window. There was no light inside. I didn't know who slept every night and awakened every morning behind the glass pane now, whose room that was. Didn't know whose life was unfolding up there. I gave the driver the directions to the house where Jack now lived.

"HEY."

And that nod of the head, just one nod downward. The little smile, mouth closed. I'd only been seeing it for fifty years and more.

He was at his doorway—the house he and Janice had built on Bexley Park Road at the end of the 1990s, the house, they told themselves, where they were going to spend the rest of their lives.

We just looked at each other. The cab had pulled away.

I looked him in the eyes and he did that nod again. Summing up everything. It had always been that way—whether we hadn't seen each other in person for fifteen minutes or six months, there were no words needed, no effort in catching up required. It was always as if we'd stopped in midsentence somewhere, and were picking it up now, and no re–summation was necessary or wanted. There would be a brief silence, like an invisible ellipsis, and then we'd be right back where we were, where we always were.

"It's raining," he said, with that startled uptick in his voice that forever had seemed to heed no inner governing mechanism to differentiate between big surprises and small, the uptick that was present whether it was the beginning of warfare he was hearing about on the evening news, or the fact that rainfall had started and he hadn't noticed it before. The uptick in his voice, from the time he was a boy, was the aural equivalent of a cartoon character's slap to his own forehead—a quick and seemingly involuntary gesture of mild shock.

"It was raining when my plane landed," I said.

"I haven't looked outside in a few hours," he said. He was wearing a baseball cap; the radiation treatments on his brain, and the chemotherapy, had begun, and his hair had started to fall out.

We walked into his living room and sat.

"Boy, Greene," he said.

"I know," I said.

And then, with his next sentence, he sounded not like a man in mourning or in terror for what had happened and for what was going to happen, he sounded like someone who had just won the World Series, because this is what he said—this is what he said before anything else:

"You should have seen Maren at the doctor's. I was so proud of her."

Maren was his daughter; she worked in the fashion industry in New York, and she had flown home to be with him, and had accompanied him and Janice to meet with his physician in Columbus.

"She was sitting there with a notepad," he said, his voice full of something close to animated excitement, "and everything the doctor said about what we needed to do, and what we should expect, she was writing it all down. She was being so responsible, and asking all the right questions . . . She's turned into such a fine young woman. I sat there and looked over at her, and it really made me feel great."

That's what he felt, at a moment when he was hearing the details of a prognosis that might make another man wince, or even weep. He felt great, he said. His daughter made him feel that way.

I had no idea what to say, so I just said what I meant: "It's good to see you."

JANICE CAME INTO THE ROOM AND SAID, "So . . . big dinner tonight?"

It was going to be an ABCDJ reunion. The significance of it, or the urgency, did not need to be stated.

"You coming?" I said. I knew she wasn't.

"Why, you want me there, Greene?" she said. Meaning: Don't even pretend that anyone other than you guys is welcome.

She was exhausted; she virtually hadn't left Jack's side since the moment she found him passed out on the bedroom floor. She and her sister and a few women friends were going out to dinner somewhere else, to pretend to put all of this aside for a couple of hours.

Jack had switched the television set on. An anchorman was talking about events overseas. Something about a disputed treaty somewhere.

I looked over at Jack, weary beneath that baseball cap. I made myself a silent promise. Whatever was to come—if the doctors were right, if this was an inevitable march toward death, if that's how this was destined to end—it wasn't my friend's death I was going to allow myself to dwell on. His life was what I cared about right now; his life, and our friendship. The other—the bad—I couldn't do anything about. His life and our friendship—all of it, starting half a century before and stretching up until this night and however far beyond we were allowed . . .

His life, and our friendship, were it. They were everything.

"What are you looking at?" he said, catching me.

"The news," I said.

"Then look at the television and not at me," he said, knowing the truth, as always.

Six

THE DINNER WAS PROBABLY OUR THOUsandth meal together, give or take a couple hundred cheeseburgers. Yet maybe the first dinner in twenty years.

The five of us had gone our separate ways. All those summer nights when we would be crammed into Allen's blue Ford, "Pretty Woman" or "The House of the Rising Sun" or "Where Did Our Love Go" playing on the car radio (we'd reach our arms out the open windows and in unison bang our fists down against the metal roof each time the Supremes sang the first syllable of "baby": "*Ba*-by, *ba*-by, where did our love go") . . .

All of those winter weekend afternoons when, nothing else in the world to do, we would cruise (with the heater going full blast and the windshield steamed up) to the

Ranch Drive-In or the Burger Boy Food-o-rama for our third or fourth lunch of the day . . .

All of those early mornings when, wanting each other's company even though we'd been hanging out until well after midnight the night before—what could we possibly have had to say to each other that we hadn't said before going to bed, what could have happened between 2:00 A.M. and 7:00 A.M. that made it so important we gather for breakfast, too?—all those mornings when we'd straggle in one by one to the counter of the Eastmoor Drive-In, just outside the Bexley line on East Main Street, a new day beginning and us together once again, as if decreed by Ohio statute, as if there was a law on the books that we had to be together every day. . . .

ABCDJ.

Allen had been the one among us who didn't live in Bexley. His parents had a top-floor rental in what was called "the only luxury high-rise apartment building in Columbus," in those days before condominiums and co-ops. The Park Towers, as it was named, soared all of seventeen stories. A wry, wiry kid with sarcasm somewhere behind his eyes and a swagger of indeterminate psychological origin in his step, Allen was most often the wheelman—the blue Ford was his (I think his dad leased it through the hotel-and-restaurant-linen-supply company he ran). For some reason the rest of us never figured out and—hard as it is to believe—never asked him about directly, Allen's parents sent him off to military school during high school, so he was never with us as much as he and we would have liked. Not that we and he weren't together more than most human beings on this earth were intended to be together: up to twenty hours a day in the summers, probably sixteen or eighteen hours a day during school vacations and on the weekends when he would come home from West Virginia. He got screwed. That's how we summed up the fact of his uniformed existence at military school: He got screwed, being sent down there, but he was still one of us, screwed or not.

B was me.

C was Chuck, the first among us to have a Beatle haircut, which only accentuated his nothing-matters-enough-to-wipe-this-grin-off-my-face outlook on the world. His father was a businessman who'd had a lot of very high highs, and a few of the deepest lows, and if Chuck was affected by that it didn't seem to cloud his mind, which was usually busy enough trying with great effort to unravel lesser confusions. Once we had been looking for some girls who on the phone had given us driving directions on where to meet them, up on the north side of Columbus near the Ohio State University campus. Chuck—in moccasins, Bermuda shorts, no socks, a white T-shirt, a dress shirt over that, the shirt buttons unbuttoned all the way down and the shirtsleeves rolled halfway up; that's what all of us wore during the summer, every summer day, that's what I undoubtedly was wearing that day, too—was driving, and in his hand he had the directions he had written down. "I can't find Old and Tangy," he said, handing me in the shotgun seat the piece of paper. I looked at it, saw what he had written as our turnoff point—"Old and Tangy," that's what he said the girls had told him—and I joined in his puzzlement as, lost, we drove back and forth, parallel with and oblivious to the Olentangy River.

Dan had been the most indecipherable of us. A short

kid and, were it not for his small stature, a potentially great mainstream-sport athlete, he had seemed to lead an interior life that had little in common with the world the rest of humankind inhabited. His dad and grandfather owned a fish market in downtown Columbus, and Dan took great pleasure in working in the freezer—he loved the freezer, he hung out there any chance he got, I think he took his lunch hours in the freezer. All of the Dick children had names that started with D—his older brother was Dicky Dick, his older sister was Darianne Dick, his younger brothers were Donald Dick and David Dick. Dan used to mention an imaginary frog—it was a specific frog, Dan said that the frog's name was Reedeep Reeves—and Dan as a teenager was prone, out of nowhere and apropos of nothing, to make sudden pronouncements such as: "Don't tell a barber how to cut hair." In September of our junior year in high school, at the first Friday night football game of the season, Dan sat with the rest of us in the stands and, midway through the first quarter, loudly announced: "This is disgusting"—he disapproved of the way the team was playing, they were big but they weren't very good—and he stood up, walked out, and disappeared. We found him after the game, running frantic wind sprints alone back and forth across his family's front yard; he pantingly informed us that he had been doing this without a rest break for the almost two hours since he had left the stadium. He hoped to grow during the school year, he said, and when and if he did, and he became football size, he intended to be in shape for the next season.

J was Jack.

The meal was probably going to be the thousandth together for ABCDJ, give or take.

I'D MADE THE CALLS, BEFORE LEAVING Chicago for Columbus.

Allen had been at his law office up in Canton; he immediately processed what I was telling him about Jack, I could sense him deciding to cancel all his appointments and ask for continuances for his court appearances, he understood without asking more than a question or two that this took precedence. I reached Dan at the cold storage company he and one of his brothers ran on Columbus's west side; when I asked the receptionist for Dan Dick, her reply was: "Which Dan?" I didn't know what she meant, and she explained that Dan Dick Sr. and Dan Dick Jr. both worked at the company, and when I asked how old Dan Dick Sr. was and she just laughed, I said that was probably the one I wanted. I asked her to connect me with his phone. She said: "He's in the freezer."

When Dan emerged, I told him about the dinner gathering, and he said of course he would be there. As I'd left my home for the Chicago airport, I'd reached for my keys and, without thinking about it, brushed my hand against the pewter beer mug that has always sat on the shelf where I toss those keys. It's tarnished now, and most of the time I don't even notice it's there; it's filled with coins that I have pulled out of my pocket at the ends of the evenings all these years. Quarters and dimes and nickels and pennies, right up to the top. On the front, a little over halfway up the side of the mug, is the engraved inscription. We'd given the mugs to each other in September of 1965, as gifts before we left home for college and whatever else might lie ahead. It had been Jack's idea; he's the only one of us who would have thought of it, who would have known that someday we'd be glad we'd done it.

I'd looked at my watch to make certain I wasn't going to be late for the airport, and I'd grabbed the keys, and my hand had hit the old beer mug with the short engraved inscription: ABCDJ.

IT WAS POURING BY THE TIME CHUCK'S car pulled up in front of Jack's house.

Chuck had called the Top and made the reservation; of course it was going to be the Top. In the middle of the 1950s, with the fathers who lived in Bexley back from the war and well along in establishing their careers and their families and their lives in the peacetime Midwest, the Top steakhouse had opened a few blocks outside town at the intersection of East Main Street and Chesterfield Road. The owners would have built it in Bexley itself except that Bexley had been dry at the time, you couldn't legally buy a drink there, and if a man and his wife were going to have a shrimp cocktail and a New York strip and a baked potato for dinner, there's no way they were going to want to do it without a martini on the tablecloth. So the Top was a ninety-second drive past the Bexley border.

Ring-a-ding-ding—that was the idea. If Frank Sinatra had ever come to Bexley, which he never did, not once, the Top is where the mayor would have taken him. Black leather booths and dim lighting and the feel of imminent action—well, at least more of the feel of imminent action than you were going to get at Ralph and Jim's Barber Shop, around the corner, or at Swan Drycleaners, just down the street—the Top was there for a reason: to give the men and women of Bexley a place to step into Vegas or Manhattan or Chicago for a couple of hours, to sense short-term snazziness, and then, when they came back outside onto the Main Street of Midas Muffler and Chicken Delight and the squat branch office of the Ohio National Bank, to be speedily and safely home to their children—to us—before the streetlights had been on for very long.

We began going to the Top ourselves when we were eighteen or so, taking our dates there on important nights and sitting in the same booths where our parents had first sat, holding the same tall menus in our hands, eating sirloins in the dimness, feeling somehow more worldly just by being inside those walls and not having the host eject us on the principle that we were merely us and therefore had no business in such a swanky establishment. The snazziness of the Top, it turned out, was not so short-term after all, because here all of us were, all these years later, taking Jack out for the evening, and of course it was going to be the Top, where else would it be?

Chuck's car was on the street outside Jack's house and with the rain beating down I walked Jack to the car, holding an umbrella over his head, hoping that he wouldn't notice I was trying to shield him, knowing that he would.

AND THEN THERE WE WERE, AS IF NOTHING had changed.

The Top had saved us a long table toward the back—it's not a very spacious restaurant, it tends to fill up, but they'd set up a table close to the fireplace, in which logs were burning even though spring had begun. Look at us.

That's all I could think in those first moments. The boys from that blue Ford hardtop, the boys of ceaseless nights on summertime streets . . . look at us.

Allen had driven straight down from his law practice on the northern edge of the state; he was dressed for court: dark suit, expensive patterned tie, blue shirt with contrasting white collar. He carried himself with the flinty confidence of a trial attorney who, having faced a career's worth of adversaries-for-hire, could walk into any courtroom in any of Ohio's eighty-eight counties to present opening arguments without his pulse rate jumping even a beat. The face of the sardonic kid had turned into the face of the bulletproof professional advocate, and the teenager's swagger had simply been a preview of how he moved through a room now.

Dan, in a dress shirt open at the collar and no tie, was lined and gray only if you looked closely at his face and made yourself catalog the details of its landscape; something about him was still so boyish, still so unguarded and shambling and free from guile, that anyone who had known him forty years before would see not the weathered man who sat with us now, but the eternal kid who resided inside his body. Reedeep Reeves, that frog who lived solely in Dan's imagination, seemed to sit invisibly on his shoulder. Dan thanked me twice for thinking to include him in the dinner; I didn't understand why he would say such a thing—the dinner was not plausible without him, it would feel incomplete—but he had been the one we had seen the least of in recent years, and he said he felt a little guilty for not having been around more. It was as if he somehow thought we could ever forget about him.

Chuck, in a black T-shirt and jeans, had lost his hair but not his Howdy Doody grin; I'd seen him so much over the years that for me the physical changes in him had been gradual enough that they seemed always to have been there. One knee was all but gone, the other was a question mark, his eyes were as brown and bright as ever but my guess was that he could use a hearing aid, his *huhs* were becoming increasingly frequent. He had a glass of vodka in front of him, as was usual at the dinner hour; his smile grew wider as he looked around the table at all of us, although I couldn't be certain if it was a smile of contentment to be with friends or merely the cocktail beginning to do its work.

Jack had on three layers of clothing—a dark undershirt, a dress shirt and a zip-up sweater. The treatments he was enduring had made him susceptible to temperature changes, he told me; his thermostat, as he referred to it, was all fouled up, and he was cold a lot. He'd asked if I thought the owners of the restaurant would mind if he kept the baseball cap on, and I said I could not imagine they would object—word spreads pretty quickly in Bexley, everyone knew about the radiation treatments and the chemotherapy. He'd always had this little gap between his top front teeth, one of them was chipped or something, and it had perpetually had the effect of making him seem unthreatening, a little vulnerable. Whenever anyone saw photos of the five of us when we were boys, they would invariably say the same thing when they got to Jack's picture: "He's the nice one, isn't he?"

And there we were. We talked very little about the specifics of what Jack was going through—he shot us a glance early in the evening that told us: Don't—and instead we laughed as much as we could, and he laughed with us. It started with my socks. They'd gotten soaked in the rain, somewhere between the front door of Jack's house and the front door of the Top, and when I looked down to see if I might take them off and go sockless, everyone took note of the fact that they were full of holes. Allen said it was pathetic—he tossed a dollar on the table to start a fund to buy me new socks, and then Chuck and Dan tossed dollars, and Jack reached into his pocket and he put a dollar on the table, too. Not exactly the stuff of top-drawer comedy, but on this night seeing Jack light up in a smile, seeing him laugh with the others, was like a gift, for the first time in so many days he was thinking about something other than what had fallen upon him. We told stories about times half-forgotten, Jack showing something close to glee at some of the memories, and at one point I brought up the chicken at the Latin Club banquet.

Actually, what I brought up was about an Internet search engine—something, when we were first friends, no one could have envisioned. I said that I had been fooling around on the computer one recent night, and had put Dan's name in a search engine, just to see if anything would come up.

What had come up was this:

Will the following students please report to the principal's office: Dan Dick. Bob Greene.

That's what the computer had returned to me. In 1964, Dan and I had been on the varsity tennis team, and one day after practice we had come into the locker room—to find boxes and boxes of individual fried chicken dinners from Willard's Restaurant piled up on one of the wooden benches. The school's Latin Club was going to have its annual banquet in the gym that night, and the chicken dinners, for some reason or other, had been left for safekeeping in the locker room.

Dan and I had looked at each other. We were hungry after practice. So . . .

We figured no one would miss a couple of the dinners. We ate them. And soon enough, during homeroom period, came the announcement on the school's loudspeaker system, piped into every classroom in the building: "Will the following students please report . . ."

Opal Wylie, the school's Latin teacher and adviser to the Latin Club, had become distraught when, at the banquet, the chicken dinners came up two short. She had gone to C. W. Jones, the school principal; he had launched an investigation. So there we were, Dan and I, on that long-ago morning, emerging from the doors of our separate homerooms, meeting in the empty first-floor hallway, walking down toward C.W.'s office and certain punishment. . . .

I had written about it at some point in the previous twenty or thirty years, and along came the Internet to suck up just about every piece of public writing in existence—and when you put "Dan Dick" into one of the popular search engines, what came back to you on your computer screen was: *Will the following students please report to the principal's office: Dan Dick. Bob Greene.*

I thought this was amazing, this was delirious, our theft of the Latin Club chicken was now available to any computer user in any corner of the globe, and I was telling the story of the computer search to Dan and to the others at the table at the Top.

And Dan said:

"I didn't steal the chicken."

I said: "What are you *talking* about?"

He looked like a person accused of something and trying to bluff his way out of it. "I didn't steal any chicken," he repeated.

"Dan, that's not the point of the story," I said. "We stole the chicken. It was forty years ago. The point of the story is the Internet."

He paused for a moment. "Oh," he said. "Oh. OK."

Jack was laughing so hard I thought tears might come down his cheeks. Allen said, "The point of the story isn't the chicken, Dan. Everyone always knew you stole the chicken." Dan said, "Well, I don't know why it would be on Greene's computer," and Chuck called for another drink, and I was so grateful for this, so grateful for the laughter that was filling the night.

Chuck's daughter arrived with a camera in her hand. Alyssa Shenk was twenty-eight; she knew where we were going to be on this evening, and she wanted to take a picture so we all would have the memory. Another one of our oldest friends—a guy named Pongi—had joined us at the dinner table, and Alyssa had all of us stand up and pose against a side wall of the dining room.

The very idea of this, back when we first knew each other, would not have seemed conceivable: Chuck would have a daughter? A grown woman? And she would be directing us where to stand in a picture someday? His daughter would be taking our picture at the Top? But that's what had come to pass, and we put our arms over each other's shoulders and Alyssa told us to smile, and as she snapped and the flash went off, I asked myself: Are the people in this picture—are we—what fifty-seven-year-old men were supposed to be like, back when our dads were the ones who were fifty-seven at the Top? Are the things that fill our hearts, the things that make us who we are, similar in any way to the things that filled our fathers' hearts, that made our fathers laugh? Are we what they were like at fifty-seven—and if they were like we are tonight, why didn't we see it? They seemed so much older, back then—their fifty-seven seemed to be such a stern and calcified fifty-seven, at least through our young eyes.

That was probably it—our view of our dads, when they were the age that we were now, probably had more to do with our young eyes than with the realities of our fathers' lives. We draped our arms over each other, and Alyssa took a few more shots just to be safe, and when the enlargements came back and were mailed to us several weeks later, I looked into the faces of the men in the photograph—looked at us—and what I saw were without dispute the faces of men who were fifty-seven, not a day younger. That's who we were—that's who our dads had been. It was right there in the photo.

And I knew one thing:

Whoever was to see this picture, no matter when, whatever stranger might someday gaze upon it, that stranger would look at the men smiling together in front of the restaurant wall, would single out the man in the center, the one bundled up in the T-shirt and the regular shirt and the sweater, the one with the gap between his two top front teeth, the one who looked a little tired behind his smile—the stranger would single out Jack. And would say:

"He's the nice one, isn't he?"

THE CONCRETE-AND-STONE WALK LEADING from the sidewalk up to Jack's house had a big crack in it.

I noticed it as we were hurrying through the rain to his front door after dinner. All of us had come back home with him; I saw the crack, and I knew how careful Jack always was about keeping his property maintained, any of his property—no socks with holes in them for him—and my first thought was that he probably had not been feeling well enough to deal with repairing it. My second thought was worse. It was whether he ever would. Whether it would be a task, down the line, that fell to Janice, alone in the house. It wasn't a thought I wanted to linger.

Which it didn't, because Janice was inside when we got there, making sure the place was bright and welcoming; she was asking how our dinner had been, and motioning us to couches and chairs in the living room, and for the next few hours we did our best—which wasn't much of an effort, all of us wanted it with all our hearts—to turn that living room into an old blue Ford, to turn the stormy night outside the windows of Jack's house into a cloudless June noonday sky, to put ourselves back in a place and a time when any trouble we might encounter was mitigated by the fact that we were encountering it together, when trouble was accompanied by the voices of the Beach Boys or the Four Tops coming out of WCOL on the car radio, when all we needed to do to outrun those troubles was step on the gas pedal and go a little faster.

We sat in Jack's house and we didn't want to leave, no one wanted to go home, and any transcript of what we said would probably read goofily, but the words weren't what mattered, what mattered was the laughter. Allen was staying at the Hyatt downtown; I'd remembered that once when he was staying there he had said something about getting a government rate on the room because a court case he was handling in Columbus had something to do with government business, so now I said to him,

"You getting that government rate this trip? You wear your old military school uniform when you checked in?"

He shot me a look and shook his head in disdain, and Chuck said to him, "You probably did wear your military school uniform and say you were in the Army." Jack and Dan started laughing, and I played the parts both of Allen and the desk clerk. "You're walking up to the front desk in that old green uniform and the sleeves are halfway up your arms and the cuffs are six inches above your shoes," I said, having trouble breathing as I tried to talk through my own laughter. "You've got your military school cap on. And the desk clerk"—now I was out of control, I was unashamedly breaking myself up, I loved this moment and I loved the sight of Jack, over on a chair, laughing, too, wiping his eyes—"the desk clerk recognizes you and salutes you, and"—here I saluted—"you salute her back, and she says, 'Welcome back, Colonel Schulman, your government rate as usual?'"

"I don't know what you're talking about," Allen said, doing his best not to laugh himself.

"Thank you for your service to our nation, Colonel Schulman," Chuck said, saluting in Allen's direction, imitating the desk clerk. "Is there anything we can do for you during your stay, Colonel?" Jack was doubled over, and it wasn't from pain, I wished Chuck's daughter had come back to the house with her camera, because it's a sight I would like to have preserved forever, the sight of the pure joy on Jack's face, and then Dan stood up and—as if this was somehow part of the conversation, as if what he was about to say was the next logical thought—he announced:

"There was a frog night at this church near my house last weekend."

Chuck's face was crimson, I thought he was going to fall to the floor he was howling so hard, and he said to Dan: *"What?"*

"They have them every once in a while," Dan said, all seriousness, presumably referring to the frog nights, which my guess was had nothing to do with Reedeep Reeves, but then, who really knew?

"What are you talking about?" Chuck, barely able to gasp out the words, said.

"It helps keep the kids off the street," Dan said, not cracking a smile.

"Dan, were you even listening to our conversation?" Allen said to him.

Dan turned to him and saluted. "That's enough out of you, Colonel Schulman," he said, and the laughter, so much laughter, filled the little room, and I know this may be hard to understand, but the laughter, the rising blend of all that laughter, sounded something like a prayer.

Seven

ALLEN GAVE ME A RIDE BACK TO MY hotel by the airport that night. His plan had been to return to Canton, and to work, the next morning. Both of us were thinking the same thing; we sat in his car in front of the hotel for a few seconds, neither of us speaking, and he said, "You know, I'm going to stay another day." I said, "I was going to ask you if you could."

We made plans to go see Jack in the morning, and he headed downtown to the Hyatt. I lay awake staring at the ceiling, and at some point after midnight the phone on the bedside table rang. I picked it up. "I can't sleep," Allen said.

I said I knew.

"This is a little too real," he said, and I said it was good he was staying over. We said we'd meet at Jack's in the morning.

BUT WHEN WE GOT THERE JACK WAS TOO worn out to visit with us for more than a few minutes; the treatments, along with the big evening out the night before, had combined to deplete him of energy. Chuck had invited all of us to dinner at his house, and Jack said he was going to take a nap because he was looking forward to that and wanted to be rested for it. We said we'd see him there.

Allen and I went back to the Top at sundown, to sit at the bar and just talk together. It occurred to me, for all the times I'd eaten in the restaurant, I had never in my life done this: had a drink at the bar of the Top. The bartender was an older guy who someone said had been working there almost since the restaurant opened.

"I wonder if he's the 151-proof guy," I said to Allen.

"Could be," he said.

I much preferred thinking about our life as it had been on the 151-proof night than the way it was tonight, and so, lost in my thoughts, I looked out of the side of my eye at the bartender, remembering the evening we'd called the Top bar to speak to the man in charge. I held back a smile.

We'd stolen some liquor. I don't want to give the impression that we were always stealing stuff; the Latin Club chicken and the bottle of liquor were the only times. It was winter of our senior year in high school, and Dan, Chuck and I had been cruising around Bexley. We'd become fans of a television series called *The Rogues*, starring David Niven, Gig Young and Charles Boyer; the show was about a charming band of con men who were always pulling off elegant heists, and that night we wanted to act like the Rogues. Just to see if we could do something the way they would do it.

So we came up with a Rogues-like plan, or as close to one as three seventeen-year-olds in central Ohio could devise. We'd look for cars in front of houses; if we saw a lot of cars, we would assume there was a pretty good chance some adults were having a party. Dan and I were wearing our varsity letter jackets we'd earned on the tennis team; we figured we'd ring on doorbells, and pretend we'd come to the wrong house by mistake, and if the people at the party were drunk enough the men would see our letter jackets, start talking about Bexley sports with us, and while the people at the door were distracted by our sports stories one of us would seek out the liquor cabinet and swipe a bottle.

I know—not the most admirable idea in the world. And after we pulled it off, I felt quite ashamed. It took us three tries at three houses, but at the third house a party was in full swing, and it unfolded just the way we had laid it out. I don't know what we would have done if we'd been caught on the way out of the house. But we weren't.

The next day, remorseful enough as it was, we discovered that we had a problem quite separate from the moral one. The bottle we had sneaked out of the party was an almost-full fifth of 151-proof rum. And the 151-proof part of it made us nervous.

That sounded like a lot of alcohol content. That sounded like something we might be right to be scared of. The beer we drank when we could get someone to buy it for us was 3.2 percent. And this purloined bottle of rum said 151 proof, right on the label. We couldn't give it back—what were we supposed to do, take it up to the door of the house we'd talked our way into, and say, "Sorry, this somehow got stuck beneath our letter jackets last night"? But we were skittish about drinking it.

That next night we sat around staring at the bottle. It might as well have had a sticker with a skull and crossbones. We each wanted a rum and Coke—we'd heard of rum and Coke, we were intrigued by the concept—but we were afraid we might be pouring ourselves an arsenic and Coke. Some rogues.

"Maybe you're not supposed to drink this stuff like regular liquor," I had said. "If it's 151 proof, maybe too much of a dose could kill you."

Chuck got the idea that we should call the Top. Where else would they know everything about liquor? But we couldn't say who we were. Chances were, our parents were having dinner there.

I made the call. When the bartender came to the phone, I lowered my voice as deep as it would go, and spoke very slowly.

"I have just purchased a bottle of rum," I said, endeavoring to sound like Charlton Heston in a Bible film, trying to transform myself into a blasé old baritone. "I happened to notice that it is 151 proof. Could you tell me how to mix a rum and Coke with it?"

The bartender, laughing, had said: "You're asking me how to make a rum and Coke?"

"Well," I said, my voice, I hoped, a voice that could part the Red Sea (or at least the Olentangy River), "I have made many rum and Cokes with regular rum. But I need to know if you make it differently with a hundred-fifty-one-proof rum."

"Put some ice in a glass," the bartender had said in the winter of 1964. "Pour a shot of rum into the glass. Then fill the rest of the glass with Coke."

"And I don't have to be careful about how much rum I use?" I had asked. "I mean, because it's a hundred-fifty-one-proof?"

The bartender, laughing anew over the phone line, had said: "I think you'll be all right."

I looked at the bartender now. He was old enough; if he'd been in his early twenties in 1964, he could be the same guy.

"You gentlemen want another?" he said, not much life in his voice, working one more early-evening shift.

I was going to bring it up to him but I didn't. I was going to ask him if it was possible he'd been on duty the night we worried about a shot of rum being strong enough to kill us. Somehow, though, a story with potential death as a punch line, even as a preposterous punch line, was not one I felt like talking about out loud on this particular night. Instead I turned to Allen and said: "You know, our dads always came here, but Jack's dad never did."

IRVIN ROTH WASN'T LIKE THE OTHER fathers.

Our dads seemed like office dads—men who put on shirts and ties to go to work, men who kept conventional business hours, men who, in their varying ways, could have been characters in *The Man in the Gray Flannel Suit*. They were of that generation—the one that came home from World War II eager to shuck their military uniforms and their bone-wearying existence on the battlefields, and embrace, if they could, a more refined kind of life. It was something to aspire to.

We'd never seen them the old way—never seen them

before they were the dads in the starched white shirts and the fedoras, pulling out of the driveway to go to the office just before 9:00 A.M., home a little after 5:00 so we could gather at the dinner table. That was our definition of them.

Irvin Roth wasn't a man in a suit, gray flannel or otherwise, not when he went to work. A compact, dark-haired fellow who seemed old beyond his years, with thick, muscled arms and a quiet voice with an underlay of gravel to it, Jack's dad was already at his fruit-and-vegetable business well before the sun was up each day, when Columbus was still bathed in darkness. It was not a genteel business—he was the middleman between the farms of the rolling Ohio hills and the restaurants, hotels and dining rooms of our part of the state, and the profit margin could be ulcer-producingly thin, the competition could be agitated and fierce. It was his show—he owned it and he ran it, and if he wasn't overseeing the loud and hectic operation at sunrise, keeping an eye on the sacks and boxes of apples and potatoes and onions coming in, marking up the ledger sheets, making sure the right shipments were loaded onto the right trucks to get to their eventual destinations on time, then it wasn't going to get done.

My memory of him is of a man endlessly tired. By nightfall there was always fatigue in his face and in his eyes. Allen Schulman's father was a dapper guy in Continental suits and white-on-white shirts, he could have passed as George Raft in a black-and-white movie, or at least as Ben Gazzara—he had that movie star aura, he looked as if he could move in fast company. Allen's dad, or so we always heard, was no stranger to the bar at the Top, having a highball at the start of the evening, celebrating sundown, not yet ready to close the books on the day's possibilities. My dad would sometimes join him—they'd been friends since they were boys in Akron. Jack's dad was never a part of it. Our dads never had a drink with Irvin Roth in their lives. They barely knew him.

"His world was different from theirs," Allen said to me as we sat in the same place our fathers had sat half a century before.

"Remember on Saturday afternoons, when Jack always seemed to be gone?" I said.

He'd be at the market; he'd be helping his father. For a while on weekends when we were eleven or twelve, some of us would ride the bus downtown, to Old Memorial Hall on East Broad Street. A local Chevrolet dealer named Lex Mayer sponsored a television program called *Lex's Live Wrestling*. It was broadcast Saturday afternoons, and it was mainly a means for Lex to sell his used Chevrolets—Buddy "Nature Boy" Rogers would wrestle Frankie Talaber or Fritz Von Goering, Handsome Johnny Barend and the Magnificent Maurice would go up against Sweet Daddy Siki and Oyama Kato in a tag team match, but the underlying purpose of the exercise would be to get each fall completed as quickly as possible so the cameras could swing away from the ring and down to the floor beneath, where Lex, a rotund, garrulous man with thick eyeglasses, would be standing next to a Chevy with some miles on it, and would announce with a flourish what a great bargain this chariot of the highway would be for some wise viewer.

That's why the show was on the air—to display Lex's cars to the viewers at home. Lex would call the action in the ring, too, he was the main blow-by-blow announcer, but his passion lay in moving the cars. Although the audience Lex cared about was the audience at home, he needed an audience inside Old Memorial Hall, too, to provide clamor and background visuals for the broadcast. Tickets were fifty cents, and for a few years when we were young some of us would be in the seats just about every Saturday. It was pretty exhilarating, for a kid.

"I remember our dads were playing poker one Saturday, and I thought he was going to kill me when he got home," I said to Allen.

The men had been out at Winding Hollow Country Club, in the card room, and the TV set had been on, and in the middle of a hand one of the players—it may even have been Allen's dad—had looked up at the screen and said to my father, "Hey, isn't that Bobby?"

Lex's cameras had been panning the audience for a reaction shot during some especially gory choreographed action in the ring, action designed to cause a frenzy, and apparently my own frenzy, at age eleven or so, was being sent out over the airwaves to every set within broadcast range of WLW-C's big transmitting tower. Including the set in the men's card room at Winding Hollow.

"Is *that* how you spend your Saturdays?" my dad had loudly demanded when he returned home that night. "Is *that* what you do with your time on the weekends?" Evidently he had not been pleased with the sight of his son jumping up and down in excitement over a flying piledriver or a figure-four grapevine. There had been some ribbing, I surmised, in the poker room.

"It's fun," I replied, or something equally inspired, which, while not satisfying him, served to hasten the end of the conversation, which would not have been the case had I said what I felt like saying to him, which was: Would you rather I be playing poker and gambling?

Jack was doing neither on those Saturdays. Jack was at the fruit-and-vegetable market with his dad, who was never invited to join Winding Hollow, who was never seen at the bar of the Top, who was up before the sun every day and who often in midevening fell asleep on the couch of their family's living room, the laugh track from their television set continuing to roar as Irvin Roth was lost to the world.

ACROSS FROM ALLEN AND ME, A FELLOW customer at the bar was not lost to the world and was not oblivious to the television set. He was looking up at the screen; the evening news was on.

The report was from Iraq. It was a particularly saddening story about the death of a soldier.

The man at the bar had ordered dinner to be served to him there; the bartender placed it in front of him.

All of us watched the news report together. It was an emotional piece, one that I knew would bring tears from many viewers.

"Man," the guy across the bar from us said. "What a terrible thing."

We all concurred.

He turned to the bartender and said:

"Can I have some pepper for this salad?"

The bartender reached to hand him a pepper mill, and Allen said to me, "Well, that didn't take long. 'What a sad thing for that soldier' to 'Bring me some pepper.'"

"And that's for a man whose death makes the evening news," I said. "That's for a man who dies in combat and the whole world is told about it. If it takes two seconds to go from feeling bad for him to digging into dinner, what chance do the rest of us have for our lives to be anything but a blip?"

"Not to the people who love us," Allen said, and we left to go join Jack.

"SO YOUR BEST FRIEND'S IN THE OTHER room," Chuck said.

He'd been using the line—the "best friend" line—for years. I'd said many times that Jack had always been my best friend, and Chuck, whenever he would call my office or my house, would say to whomever answered the phone: "Tell him it's his best friend." When I would pick up the receiver, he would invariably say: "Why aren't I your best friend?" In a tone indicating that was a distinction he was just as pleased to do without.

This was the first time he'd used the line since Jack had gotten sick, and it sounded different tonight, there was something almost gentle in it. Jack was sitting in the sunroom of Chuck's house, Chuck's house being one of the most beautiful and sumptuous I'd ever seen. He'd done quite all right for himself over the years; his house was filled with world-class artwork, was set back on sprawling and meticulously groomed grounds, with a swimming pool and a tennis court and the finest of everything from fixtures to furnishings.

Jack, in his baseball cap, was on a couch between Chuck's wife, Joyce, and Janice. I still had trouble telling the twins apart—I could do a pretty good job of it when they were together, but if I came upon one of them without the other being present, I was wrong as often as I was right. It continued to strike me as so unlikely—that two out of the five of us would marry twin sisters—that my inability to discern which of them was which was overridden by the basic outlandish improbability of the situation itself.

"Sit down," Jack said, patting the seat of the chair next to him. I did. "How was your day?" I said, and he sort of raised his eyes toward the top of his face, his gesture for: Don't ask. He was just getting used to this, and so were we—the fact that his days of feeling great were over, that either the cancer or the medicine or both were always going to take something out of him, every day and every night.

"Where'd you and Allen go?" he asked, and I told him we'd had a drink, and he nodded wordlessly, the way he always would. I'd seen that silent nod of acknowledgment all my life, it was a variation of the nod with which he'd greeted me at his house the night before. I'd never thought much about the fact that he was a nodder—some people are, some people aren't—but I was doing my best to notice everything now, I didn't want any part of this, any part of him, to slip past me. Chuck and Joyce had a Beatles greatest-hits CD on their sound system, and the music played at just the right volume through discreetly hidden speakers in rooms around the house. "She Loves You" was on, and Paul McCartney's voice from 1964 was singing about how you know that can't be bad.

We were going to make the best of this, wherever it led. I could smell the food cooking in the kitchen, and Chuck and Allen came into the sunroom and we all started talking. McCartney's voice was somewhere in the background, almost drowned out by the sound of our voices, but I still could hear it:

. . . and you know you should be glad . . .

That was it. That was it, I thought. Here we were, after all the years. Don't feel mournful about it, not now. There will be time for that later. Here we were, the friendship still with us. We had that, we had it then and we had it now, and it was a gift, all right, it was one of life's true wonders.

Jack leaned close to Allen to hear something he was saying. I wanted to tell them what I'd just been thinking, but I kept the thought to myself. I looked at them, wishing I had a way to say it.

Here we were, I wanted to tell them. Here we still were.

And you know you should be glad.

DINNER WAS DELICIOUS; WE ATE AT THE long formal table in Chuck and Joyce's dining room, and everything seemed to be fine and at one point Jack turned to me and said quietly: "I need something to spice up my food."

I wasn't sure what he meant; the food was very flavorful, and even if it hadn't been, he was the last person in the world who would ever as a guest in someone's home say that something he was served was lacking.

"I just can't taste anything anymore," he said to me.

He thought it was the radiation treatments; he thought the radiation on his brain was removing his ability to know what food tasted like, although, he said, it might be the chemotherapy. "It's like the signals that go from my mouth to my brain aren't working," he said.

Everyone else was talking away, and I could tell that he didn't want to spoil their dinner with this. I asked him if he had lost his sense of taste as soon as the radiation treatments had begun.

"At first, everything tasted like tin," he said.

And now?

"Now it all tastes like mush," he said. "No matter what I eat."

"So you really want something spicy?" I said.

"Just something to make it so I can taste the food," he said.

I excused myself and went into the kitchen to look for some hot sauce.

AFTERWARD WE ALL WENT BACK TO THE sunroom. There was a lot of sound and commotion; the Shenks' dog was barking constantly, and there were overlapping layers of conversation, and the television set was turned on in the background. I saw Jack at one point just staring straight ahead.

"This too much for you?" I asked him.

"No, I'm having a good time," he said. But each time I looked over at him, he had that faraway stare.

"You think you'd like to go home?" I said.

"In a few minutes," he said. "Listen. . . ."

"Yes?" I said.

"Will you do something with me tomorrow?" he said. "Will you do me a favor?"

"What do you think?" I said.

"Maybe I'm so tired because I haven't been getting any exercise," he said. "Will you walk with me tomorrow?"

"Gladly," I said.

Eight

"You're saying that was the room?"

Jack asked me the question. I had pointed to a window on the first floor of the Cassingham Elementary School building.

"I can't be sure, but it seems like it was that one, or the one next to it," I said.

"I can't believe you were drawing a picture of a woman in a bikini," he said.

"Why can't you believe it?" I said. "Just because we were in third grade?"

"I saved you," he said.

"You definitely saved me," I said.

We were about ten minutes into our walk—that's how long it had taken us to get from his house to where we were standing now, on the sidewalk in front of the school.

He had a grin on his face; he was much more full of energy than he'd been the night before.

"Why were you drawing a woman in a bikini anyway?" he said.

"I think it was because of Rose LaRose," I said.

Rose LaRose was a stripper. Not that, in third grade, I'd ever seen her perform. There was a burlesque house in downtown Columbus—Gayety Burlesque—and the Gayety used to take little display advertisements in the three daily papers that Columbus had at the time, the *Columbus Citizen*, the *Columbus Dispatch*, and the *Ohio State Journal*. They weren't big advertisements—maybe an inch and a half or two inches high and a single column wide. For readers who were interested enough to look, the ads announced which striptease artists were performing during a given week. I was a reader who was interested enough to look.

The Gayety couldn't show much in their ads—usually just a tiny and almost certainly outdated photo of that week's stripper leaning back in a sultry pose, maybe running one of her hands through her hair. The strippers weren't nude in the ads—they customarily had pretty substantial halter-style tops on, and bottoms that contained considerably more fabric than shorts you see on the streets (or even in airport boarding lounges) today. Because the metal photo engravings of the strippers tended to stay in newspaper composing rooms for years at a time, to be used over and over, and because the pressroom machinery of American newspapers of that era usually made it look as if photographs were being printed on pieces of Wonder Bread, the strippers in the Gayety advertisements often bore as much resemblance to smudged inky fingerprints as to beautiful and enticing women.

Didn't matter to me. I was a big fan of those little Gayety ads. And Rose LaRose—a dark-haired and curvaceous touring stripper—was my favorite. I didn't care about what was on page one of the paper, but I always ripped it open to see if Rose was in town.

Which led, one day in Miss Kellstadt's third-grade class, to me opening my notebook and attempting to draw, on a lined page, a picture of a woman in a bikini. I drew the legs, I drew the bikini bottom, I drew the torso, I drew the neck . . .

And then I heard Jack.

He was making a kind of half-whistling sound; *"psst"* doesn't quite do it, but you get the idea. He made the sound again.

He was sitting next to me (in third grade we had moved back to adjacent desks after being separated the year before), and he had seen Miss Kellstadt coming down the aisle. He had also seen what was in my notebook.

I looked up, saw Jack motioning with his head toward Miss Kellstadt, and—this was a decision I had to make in an instant—drew a smokestack on the neck of the woman in the bikini.

Out of the smokestack I drew thick smoke.

So what was on that page in my notebook was the body of a woman wearing a bikini, and, where her head should have been, a big smokestack doing what smokestacks do.

Miss Kellstadt stopped at my desk. She looked down.

"What's that you're drawing?" she asked me.

"A factory," I said.

"A factory," she said.

She stood over my desk for a second or two.

Then she moved on, and did not mention it again.

"I did save you," Jack said now, on Cassingham Road,

as we looked at the building. "I don't know what she would have done if you hadn't changed the picture."

"I'd almost bet that was the room," I said, peering across the front lawn toward the window.

HE'D THOUGHT THE INACTIVITY—SITTING around the house, spending so much time in doctors' offices and hospitals—had made him sluggish. That was the reason he'd wanted to walk through town with me: to recharge himself physically. At least that was the stated reason.

As we walked together, though, I could sense something else going on. He was revisiting his life—our lives. Every corner we came to, every street we turned upon, took him someplace different, from one juncture or another of his years on earth. A movie projector could not have provided more vivid pictures. I could tell in the way he looked around. He was seeing a story.

It was a warm spring afternoon. "Boy, I don't know what was wrong with me last night," he said. "I was really wiped out."

"Second floor," I said, gesturing toward the junior high school building. "'LJR in '84.'"

"I think I really believed it could happen," he said.

IT HAD BEEN DURING THE KENNEDY-Nixon presidential campaign of 1960. We'd been in eighth grade that fall, on the second story of the school building. Some of the students wore Kennedy buttons, some wore Nixon buttons, most, I suppose, wore no political buttons at all.

But Jack had made his own. "LJR in '84."

He'd sat down and figured it out. Under the Constitution of the United States, you had to be thirty-five years old to be eligible for the presidency. The first presidential election after his thirty-fifth birthday, he had calculated, would be the election of 1984. He'd be thirty-seven that year.

He was as self-effacing a boy as I'd ever known—he just didn't show off, ever. But something had gotten into him, and he'd worn the button most of that fall. The *L* stood for Louis; he had been born Louis Jack Roth, he always disliked the "Louis," but it must have worked for him in the context of a political slogan, because "LJR in '84" it was. Wearing that button in 1960.

"Why'd you do it?" I asked him now.

"I think I was just really excited about the presidential campaign," he said. "There's probably a time when you believe what they tell you about any kid being able to grow up to be president. I guess I thought I was giving myself an early start."

"Did other kids ask you about it?" I said.

"Mostly about the ' '84' part," he said. "I had to explain to them that 1984 was my first eligible year."

"And you got that letter from Rockefeller, too," I said.

His smile widened. "I'd forgotten that," he said.

I hadn't. Around the same time as "LJR in '84," he had written a letter to Nelson Rockefeller of New York. He'd either read about Rockefeller in a magazine or seen him on television, but in his bedroom on Ardmore Road he'd written a letter to Rockefeller saying that he hoped to go into politics someday, and Rockefeller had written back, wishing him luck.

"It was always on the refrigerator door," I said to him.

"My mom told me I should put it there," he said, falling silent for just a beat before continuing. The Rockefeller letter would have been about two years before her death.

"She said it would be good for everyone who came into our house to see it," he said.

"I remember the stationery being kind of embossed, or engraved," I said. "A dark blue official seal, I think, maybe some gold. Was he governor of New York at that time?"

"Right on our refrigerator door," Jack said, marveling now at the thought: a letter from a Rockefeller, starting out "Dear Jack," in the cramped kitchen of Mildred Roth.

"I WISH I COULD RUN EVERY DAY," HE said.

He was looking up the street, at the route he knew so well—the running route he would follow after work and on weekend mornings.

It's something he had done for so long without even thinking of it in terms of being one of life's pleasures. It was just a part of his day. But now that it was gone—now that he was too worn out to do it . . .

"Now, there's something our dads never did," I told him.

He laughed at the very thought. Irvin Roth, or Robert Greene Sr., dashing through the streets of Bexley in running clothes? Not any more likely a sight than the two of them wearing robbers' masks and holding up the Buckeye Federal Savings and Loan. Sticking up the bank would never have occurred to our fathers, and neither would going out for a brisk run.

"I think it was just a more sedentary time," Jack said.

"Once you became an adult, you didn't run. You pretty much sat still."

"Well, men played golf, or bowled," I said.

"Those were games, though," he said. "I don't think I ever saw one father or mother going out for a run when we were kids."

"If anyone had, they'd have been called 'health nuts,'" I said.

"Or someone would have called the police," Jack said. "Any grown man seen running down the street, people would assume he was running from something."

He stood and gazed once more at the road upon which he so recently had taken his daily runs.

"Anyway . . . ," he said.

I don't think he'd gone for those runs only for the exercise. I think it was that he wanted to look at the town every day. He wanted to run through a place that he loved.

THE BASEBALL DIAMOND WHERE GARY Herwald hit the home run was just around the corner, and Jack wanted to go see it.

We'd been in junior high school—right around the time of the "LJR in '84" button. There had been a game on the diamond in back of the school, a regulation game against another junior high school team, and Gary Herwald, a classmate of ours, a good-natured kid with curly dark blond hair who everyone liked, came up to the plate.

He faced the opposing pitcher and he swung the bat and connected, and the ball flew toward the left fielder, who ran back, and back . . .

And the ball dropped over the tall chain-link fence

that separated the junior high school diamond from the high school athletic fields. Gary had knocked the ball over the fence.

This had never happened before. None of us—no one we knew—in a real game had ever hit a ball over a fence for a home run. We'd all been equal, sort of—equal in the kids' fraternity of boys who had never hit a home run—and then Gary had changed all that. He had separated himself from the pack. He was a home run hitter now—he had entered that new realm of success, he was someone different by the time he reached home plate. He was there, and none of the rest of us were.

"Were you jealous of him that day?" I asked Jack as we reached the diamond, empty now.

"Of course," he said. "Everyone was. I was happy for Gary, but I wished it was me."

It was the first time we had experienced something like that. Later, in the adult world of business and gnawing ambition, we—all of us, everyone who is thrust into that larger and colder world—would go through it time and time again: seeing someone move ahead of us, seeing someone achieve something or be given something that the rest of us can only yearn for. You feel it in your stomach. You sense the sands shifting. Someone has moved beyond you, and you are witness. Someone has become something different—something better—than what he or you had been before. And all you can do is watch it happen.

"I asked Gary about it once," I told Jack. Gary was in sales for a company up in Cleveland, and I would run into him every few years.

"What did he say?" Jack asked.

"I asked him what he was thinking about as he was running around the bases after the ball went over the fence. And Gary said, 'Probably that I'd never be able to do it again.' "

"That's Gary," Jack said. "He would say that."

"He said he meant it," I said. "He said he never hit another home run in his life."

"I never even hit one," Jack said. "I don't know what's worse—never hitting a home run, or hitting one and knowing that it's never going to happen again."

We stood near home plate and looked out at the diamond, and at the fence that at the time had seemed so far away, on a spring day like this one.

HE WANTED TO TRY THE TRACK. IT WAS A part of his after-work running routine—when he would hit the streets of Bexley he would at some point run onto the high school track through the open gate just off Fair Avenue, circle it a few times before running right out the gate and back to the streets—and today he told me he'd like to go there.

So we did. No running this afternoon—just a slow counterclockwise walk, passing by the home grandstand, continuing beneath the football scoreboard, on past the visiting team bleachers. There were a number of people jogging resolutely around the track, and Jack and I stayed close to the outside edge, so as not to get in their way.

"Did you ever see the brick?" he asked me.

"You told me it was there, but I never have seen it," I said.

"Let's just walk the track one time around, and I'll show you," he said.

We found an opening in the fence at the northwest

turn of the track, and Jack led the way to a plaza in front of the high school building. The plaza had been constructed in recent years, on a place where before there had been only grass; its presence was intended as a fund-raising effort and as a way to help preserve the heritage of the town.

The bricks that formed the floor of the outdoor plaza were carved with names of generations of students who had gone to the school. The opportunity to have one's name carved into a brick was for sale; I think the price was fifty dollars to have a name cut into one of the smaller bricks, with larger stones costing more. Usually a brick contained one person's name, beneath which was his or her graduation year; in some cases entire families signed up together.

Jack had done what he'd done on his own. He had told us about it only when the brick was already in place.

"It's right up here," he said now. "They really put it in a good location. Right near the front sidewalk."

I followed him. He walked on the plaza—many bricks had names on them, many others did not; this was by definition a work in progress. The idea was to keep adding names for the next hundred years and beyond.

"Here," he said.

I looked down. It was the only brick with an inscription that wasn't a person's name, or a family name.

ABCDJ
1965

"I just thought it would be nice," he said.

"It is," I said.

"Something to show that we were once here," he said.

WE WERE ON OUR WAY TO HIS OLD HOUSE, and he said, "I think this is almost the exact place where Jerry Hockman said hello to us."

Every step he took, every direction he looked, he was finding something. He was touring his past, he was an archaeologist on a deadline not of his own making, excavating long-lost joy. I was probably the right person for him to have with him; I was the one person in the world who wouldn't have to ask him about his references, including the Jerry Hockman reference.

It's funny how a kind gesture from someone can stay with you—how the smallest choice a person makes can resonate over the years. The reverberations of cruelty and gratuitous meanness, we often hear about—absence of mercy tends to make the history books. Yet it can work the other way. Fleeting moments of kindness can echo forever.

We were little elementary school kids walking down this same sidewalk when, coming from the other direction, we saw Jerry Hockman. Jerry Hockman, that year, was the high school's star athlete—older than us, living in a different solar system than us, accustomed to hearing cheers. He had no idea who we were. We were young, invisible.

So here came Jerry Hockman, in his blue varsity letter jacket with the white *B* on the chest, and to us it was like we were seeing Johnny Unitas, to us it was like we were seeing Mickey Mantle. He had that kind of celebrity, in that town, in that year.

And our paths were about to cross.

What were we supposed to do, at a moment like this?

Get off the sidewalk, to let him pass? Offer him words of praise? Ask for his autograph? We hadn't planned this encounter—what were the rules for such an occasion?

What we did was look down at the sidewalk and avoid his gaze. What we did was fall silent and feel small.

What Jerry Hockman did was speak to us. "Hi, guys," he'd said. Just that—he acknowledged we were alive. We looked up and he gave us a smile and a nod of his head as he walked past. Tiny choice on his part—ignore the two kids or make them feel special. Tiny choice—and here, at fifty-seven years old, Jack was remembering it.

"We had to ask ourselves if it had really happened," I said. "We had to make sure we each had heard it right—that Jerry Hockman had really spoken to us."

"The next day in school, everyone thought we were lying," he said. "They thought we were making it up."

"If you'd had to choose only one of the two things—the letter from Nelson Rockefeller or the hello from Jerry Hockman—which would you have chosen?" I said.

"I didn't have to choose," he said. "That was the great thing."

"But if you'd had to," I said, loving these moments, never wanting this to end. "If someone had made you take one of the two."

"Definitely Jerry Hockman saying hello," he said. "We were younger. It was more unbelievable."

NEXT TO AUDIE MURPHY HILL—ON THE strip of sidewalk between Elm Avenue and Jack's old front lawn—we debated whether to walk onto the property.

"I don't think they'd mind," I said.

"Yeah, but if they look out the window and see two guys walking around their house, it might scare them," he said.

"We're not all that scary," I said.

"Still, I don't know who lives here now," he said. "I think we should just look from here."

So we did. The house, for certain, appeared older—it had been more than fifty years since I'd first come here, since we'd pretended to be Audie Murphy as we hunkered close to the grass and, stick-rifles in hand, crawled up that soft little slope in pursuit of enemy soldiers. We walked east a few feet until we could see the back yard, and I asked Jack: "What was that wooden thing for?"

"I have no idea," he said. "I don't think my parents ever opened it."

It was this strange-looking wooden set of doors—horizontal doors—that extended from the house and into the back yard. They were constructed to open upward—beneath them must have been steps that led to the basement. They always seemed sort of spooky to me. Like trapdoors, literally.

"They must have been here when my parents first moved in, but we didn't touch them," Jack said.

Above them and to the left was the window to the breakfast room.

"Your sister Helen really confused me in there when she talked about Chuck Berry," I said.

Jack laughed out loud—he was doing that a lot today, it was such a great sound to be hearing. "'Drop the coin right into the slot,'" he said.

The Chuck Berry song "School Days" was out and a big hit on the charts. This was 1957; we were ten. Chuck Berry seemed to be this mysterious and vaguely dangerous figure, a visitor to our ears from a wider world full of potential pleasures and pitfalls the nature of which we could only guess at; his voice had that knowing, nasal curl to it, a little like a leer, and one day the kitchen radio in Jack's house was on and we were eating Toll House cookies and drinking milk in the breakfast room. Chuck Berry was singing about "right to the juke joint you go in," and he spat out the famous line:

"Drop the coin right into the slot. . . ."

He rolled the start of the first word: "Drrrop . . ."

Helen Roth, two years older than Jack, had come into the breakfast room and said to us: "You know what he's really singing about, don't you?"

We had looked up, blank.

"A jukebox," I had said.

"Obviously a jukebox," Helen had said. "But do you know what he's really singing about?"

We just looked at her.

"A woman," Helen had said.

This was the last thing we expected to hear. This staggered us.

"The slot is a woman," Helen had said.

I have no idea whether she was right. I still think that the line was about a jukebox, and only about a jukebox. But on that day it had been such an unanticipated thing to hear, so salacious and disorienting—that Chuck Berry might have been singing a story about a jukebox but really sneaking a reference to sex into the air inside the Roths' house . . .

"The first time I ever flew first class was because of Chuck Berry," I said.

"I know," Jack said. "You called me afterwards."

I had been a beginning reporter in Chicago, assigned to a story in St. Louis, and I'd gone out to O'Hare for the flight to Missouri. Among the other passengers waiting in the TWA boarding area, his guitar case propped up against the seat next to him, was Chuck Berry. It was like seeing Abraham Lincoln.

When the gate agent arrived to begin check-in, I followed Berry to the desk. He had a first-class ticket. So, on an impulse, I paid to change my ticket to first class, too. I just had to ride in the same cabin as Chuck Berry.

I didn't say a word to him the whole flight. I was seated in the row behind him; I mainly watched him and the flight attendants flirt with each other. When the plane landed in St. Louis, I walked into the airport and went straight to a pay phone and excitedly called Jack.

"Did you find out what 'drop the coin right into the slot' means?" he had asked, as I knew he would.

"I didn't," I had said. "I was too nervous to talk to him. I wanted to ask him, but I just couldn't."

So it remained a mystery, locked forever inside that breakfast room on Ardmore Road. We looked toward the house—toward where that other mystery had been, the wooden set of double doors built parallel to the ground, more ominous and ambiguous even than a Chuck Berry lyric, something seemingly quite conventional, most mundane. But you never can tell.

"MY GYROSCOPE JUST SEEMS WRONG," HE said.

He pointed toward his feet.

"It's like sometimes I have trouble keeping my steps together."

"Your gyroscope, your thermostat . . . ," I said.

"I know," he said.

"It's got to be part of the radiation treatments," I said.

"The plate from the Toddle House is great," he said, steering us back to something sunnier. "I can't believe you found one."

"Speaking of which, do you feel like going over to where it used to be?" I said. For that journey, no gyroscope or compass would be needed. We could get there in our sleep.

Nine

If you'd had a pocket calculator—not that anyone had pocket calculators back then, this would have required one of those mammoth Univac computers that filled entire rooms, spitting out elongated cards with punch holes in them . . .

But if you'd somehow had a pocket calculator, and had added up the total number of hours that the five of us had spent together in any of a hundred specific places, including visits to each other's homes, the winner by a landslide would have been the Toddle House.

I don't even know how it managed to stay open—there were no tables, no booths, just the counter with maybe a dozen stools. It ran twenty-four hours a day, every day of the year including Christmas, which means someone had to pay three shifts of short-order cooks each day. The cook was usually the only person on duty; it still is a puzzle to me how the revenues that came in from twelve stools were sufficient to keep the Toddle House alive.

To be sure, five of the stools were quite frequently accounted for by ABCDJ. We did our thinking at the Toddle House, and our brooding at the Toddle House, and our conniving at the Toddle House—we didn't even have to talk much to each other when we were there. The Toddle House to us was a little like a chapel, had any chapel in the world sold cheeseburgers, hash browns, Cokes and slices of banana cream pie.

It was right on Main Street. There were Toddle Houses all over the United States, although they're gone now. Our Toddle House was on East Main just off Remington Road, and if we couldn't find each other anywhere, we'd just head for the Toddle House. It was home base. Everyone would show up.

(There was one short-order cook/waitress at the Toddle House who invariably answered the phone with the weirdest, most wavery, quavery, high-pitched, yodel-like shriek: "Toddle House!" It's not possible to convey the sound of her voice in print—it was like something out of the Swiss Alps, it was like she was summoning an elk or something from the next mountain pass. For years afterward, whenever any of us would meet up with any of the others for a meal, at one point someone at the table would yodel at top volume: "Toddle House!" It was an anthem, that yodel.)

When I first heard the news about Jack—when I received that first phone message from Chuck—I knew what I would have to send him, and it wasn't a Hallmark card.

JACK TALKED BUSINESS ON OUR WAY TO where the Toddle House used to be. To be specific, he talked about how worried he was about what was to become of his business.

He worked for himself. After many years of being employed by various companies in the wholesale merchandise industry, he had set out on his own, packaging and selling private-label food products to retail operations around the country. I could never really understand how it worked—the ins and outs of that kind of enterprise were foreign to me. But that's what he did, and that's what he was on edge about.

"I don't know what's going to happen," he said. "I'm not even supposed to drive a car by myself while I'm getting these radiation treatments, and I hate asking Janice to drive me up to my office."

"I know it's easy for me to say," I told him as we walked toward Main Street, "but you can't think about business right now. You just have to think about getting well."

It was, in fact, too easy a thing for me to say—it was too glib. I wasn't the one who had to pay Jack's household bills, I wasn't the one who had to pay his insurance premiums, I wasn't the one who had to pay the rent on his office space. He had to balance all of those worries at the same time he was fighting the cancer inside him.

"Who ever thought we'd have to think about business?" he said.

"I guess before you leave your first world, you don't think of yourself as anything other than what you are," I said. "Imagine us sitting at the Toddle House, talking about office expenses."

"I still think about that time we saw Mike Ingram at Lazarus," Jack said.

Mike Ingram had been a guard on the Ohio State football team in the late 1950s. I believe he may have been captain his senior year. Tough-looking, brawny guy. A hero in Columbus, a gladiator in a scarlet-and-gray uniform.

One holiday season Jack and I had been at the F & R Lazarus department store downtown, and we saw, carrying a tall stack of cartons, someone who apparently was working in the Lazarus stockroom. It was Mike Ingram, post–Ohio State football.

We stared—how could we help it? And Mike Ingram stopped in his tracks, looked right back at us, and said, with bite in his tone: "Yeah, it's me." Meaning: Go ahead and stare if you must.

Couldn't really blame him. There was nothing wrong with what he was doing—he was earning some money in the stockroom. But he was out in the world now, he wasn't where he had been when everything was bathed in sunlight and free of choiceless tedium; he wasn't in Ohio Stadium, hearing the approving roar from 78,000 people who loved him and his teammates. He had moved past that first of life's roles, as everyone does. It was his misfortune to have been famous very early; there must have been dozens and dozens of men in their twenties working in the Lazarus stockroom that holiday season, but Mike Ingram was the one destined to attract curious gapers, because he was no longer who he was supposed to be.

"And Mel Nowell at the Union," I said. The same thing had happened at the Union department store in the Town and Country shopping center on East Broad Street. We'd been in there one day as teenagers, and selling ties in the men's department was Mel Nowell, who had been a starting guard on the 1960 Ohio State NCAA champion basketball team. The best college basketball team in America: Jerry Lucas, John Havlicek, Mel Nowell, Larry Siegfried and Joe Roberts. And then, not all that farther down the road, he was on the first floor of the Union wearing a shirt and tie, selling ties. For all I know this may have been part-time work for him, and the Lazarus stockroom job may have been part-time work for Mike Ingram. But the game clock in Ohio Stadium had run down to zero for Ingram, the game clock in St. John Arena had zeroed out for Nowell. It was time for Act 2. It happens to all of us.

We could see Main Street straight ahead.

"It's probably best when you're a kid that you don't know that things are going to get complicated," Jack said. "There's nothing you could do about it, anyway."

IT WAS A GOOD THING THE TODDLE HOUSE was always open on Christmas, because we were always there on Christmas (late at night, as the holiday was in its dwindling hours); we were always there on Thanksgiving (early in the morning, before our families began to gather for their turkey dinners); we were always there on New Year's Day (at noon, as the world and we were stirring to tentative wakefulness from parties the night before). We felt welcome; a lot of times we were the only company the cook had, I sensed that there were nights when, as with the driver of the Bexley bus, we helped break the loneliness of the person on duty. We were voices to fill the air.

We were at ease being silent at times in the Toddle House, yes, but when we did speak to each other we seemed on occasion to say unexpectedly private and heartfelt things, the kinds of things I don't recall us saying in other settings. I think, looking back on it now, the reason may have been this: On those side-by-side stools at the counter, we didn't see each other's faces. We looked straight forward toward the grill, or looked down as we ate our burgers and drank our Cokes, and we could talk without making eye contact. We could say certain things without being required to watch the reaction.

When I heard that Jack was sick there was something I wanted to give him, and I had no idea where to find it. But then it occurred to me that the Internet had to have some salutary reason to exist.

"IT'S SO NICE OUT TODAY, I FEEL LIKE WE should be going up to Moe Glassman's," Jack said.

That's what we would do in the late springtime; that's what we would do to get ready for summer. Moe Glassman's was a store on the Ohio State campus that sold clothing to male undergraduates. It's where we all, before we were old enough to be in college, would go to buy our moccasins. The mocs-without-socks were our definition of summertime. I don't know how it started.

"What an odd thing, all of us wearing the same kind of clothes in the summer when we were kids," I said. The old Bermudas, the white T-shirts, the dress shirts open all the way in front, the sleeves rolled up—I don't know who did it first (although it was probably Chuck), I don't know who got the idea, but that's how it always was, and each year it started again with that first hint of summer in the air, and the drive up to Moe's.

"Imagine if once you got older you dressed the same way as everyone you worked with," I said. "People looking in from the outside would think you were all nuts."

"What are you talking about?" Jack said. "That's exactly what people do when they're in their thirties and forties and fifties. Look around any business office."

I guessed he had a point. But shirts and ties and slacks didn't seem to be in the same league as the mocs-with-no-socks uniform.

"Going up to Moe Glassman's every spring sort of felt like we were joining a club," I said.

"Except we were already in it," Jack said. The old Toddle House was right around the corner.

ONE TIME WHEN JACK WAS IN CHICAGO on business we were on a street downtown when a loud argument broke out. A cab had pulled up close to the curb to let its passenger out; the passenger was taking her time, and the person driving the car behind the cab screamed: "Get out of the street, you . . ." The words that followed were obscene and ugly to the extreme. Horns sounded up and down the block, the cabdriver took the incident as a challenge and refused to move, more voices were raised, more curses were spat, and Jack and I had stood on the sidewalk thinking about how far we were from home. Even though Chicago had become my new home.

We came from a place where you could get out of a car

on virtually any street and take an hour doing it if you wanted, because almost certainly no one would be behind you, and even if someone was behind you on that particular day, he or she would have plenty of room to just pull around. We came from streets where shouting was an unusual event. There were no parking meters, even. None were needed.

Now Jack said that instead of walking the last half block up Main Street to the place where the Toddle House had been, he wanted to cut through the alley. "I'm trying to remember where we always parked," he said. "I'm thinking there was a little place in back."

So we headed west into the alley, and he had been right, there was a little paved lot behind the orange-red-brick building, with a handful of sets of painted lines extending from the back wall. You see something like that thirty or forty years later, and it washes over you, you know before you even focus that you've been here so many times, you can feel yourself steering your car between the lines and up to the bricks. Even in daylight and on foot this afternoon I could see the headlights from my car beaming against that brick wall; the sun was high in the sky and I was walking, but I could see my headlights hitting the wall in the moments just before I shut them off.

"I don't know if these lines have had a new coat of paint since we used to go here," Jack said, looking down.

It used to feel like an overture. Just driving onto this patch of blacktop behind the Toddle House used to have the quiet power of an orchestra warming up before the start of a play, as the actors gathered one more evening.

I HAD FOUND WHAT I WAS LOOKING FOR on the Internet.

In those first days after I had heard about Jack, I searched the worldwide computer network, and there it was, for sale in a family-owned antiques store in the deep South.

An old dinner plate from the Toddle House chain—a heavy china plate of the kind from which we used to eat our cheeseburgers and pie.

Before the Internet, I suppose, I never would have found it for Jack. At least not in time.

But there it was, right on the screen, a photo of it.

Around the circumference of the plate were multiple maroon imprints of the little cartoon chef, and the name of the place in that distinctive typeface of off-balance lettering, each letter at a different angle, sort of a jolly, dancing typeface. Toddling, even.

There was a slogan, too. I guess I'd forgotten it. But there it was, right under the words *Toddle House* on the dinner plate, on my computer screen.

It was about as bashful yet proud a slogan—a slogan without pretensions or imperiousness, yet confident nonetheless—as any you could come up with.

"Good As The Best."

That's what the Toddle House had promised. We may not be the best. But we're just as good as them.

It's a feeling we understood very well, in the deceptively quiescent middle of Ohio.

PIZZA PLUS.

The sign in front of the building said that was its current name, its present incarnation. Jack and I, having walked around from the parking lot in back, looked in the front window. There were only one or two customers present.

"It's been Pizza Plus for a lot of years now," Jack said.

I could see there was no dining counter and no stools; the place had been reconfigured so that pizza ovens and submarine-sandwich heaters were where the short-order grill had been. This was mainly a walk-in-and-carry-out operation, or so it appeared from the outside. The two guys who were working plainly had been born many years after all those days and nights we had spent inside the little structure. They could be excused if they had no idea of what had been here before.

"Forty years from now, the kids who come here now for pizzas and subs are going to look back on this place with the same feelings we have about the Toddle House," I said.

"I doubt it," Jack said. "It's not the same when you carry the food out. It wasn't the food we cared about. It was the sitting around together."

"Don't knock the food," I said. "The sitting around was great, but think about the Toddle House chocolate pie."

"There was something about the way the chocolate pie tasted when you had it with a Coke on ice . . . ," Jack said.

As I looked through the window something came to me—something I hadn't thought about in years—and it almost made me shudder. I knew immediately I wasn't going to mention it to Jack.

After we'd gone off to college we would still gather in the summers at the Toddle House. And late one summer night when we were nineteen or twenty, we ran into a guy with whom we'd graduated—not a close friend of ours, but someone we'd known all our lives. He was at the Toddle House with a couple of his own friends, and he was wearing a white shirt and a tie.

We all had sat at the counter. He had a summer job with a funeral parlor; his tasks included going to homes where someone had just died, and helping to carry the body out and transport it to the funeral home for preparation for burial. It was a job that required a somber demeanor while on duty.

Off duty, though, was something else. He had laughed that night at the Toddle House and had told detailed and distasteful stories about what he encountered on the job. He had described the bodies; he had mockingly imitated the family members in their first moments of grief. Part of it, I suppose, was gallows humor; part of it may have been his way to deal with that kind of darkness. And when you're young, and death has always seemed like something so far off over the horizon, maybe certain defense mechanisms kick in when you suddenly are required to look death in the face on a daily basis.

But I remember how disturbing it had been to sit there at the counter, appetite gone, and hear him, just off work for the night, making too-graphic banter about what he had seen during the course of his latest shift. It wasn't just the gruesome specifics that were so unsettling, although those specifics were bad enough, and in a restaurant, at that. It was that he was laughing—

defense mechanism or not—at the pain he had witnessed. Maybe he was just showing off. Maybe he was trying to make us think he was tougher than he really was. There was a smell of formaldehyde to his clothes.

"Let's go in," Jack said, and I said sure.

I HAD ORDERED THE DINNER PLATE FOR him.

It was easy; type his name and address into the computer, enter my credit card number, and it was done.

Within minutes I received an e-mail from someone at the antiques store in the South; apparently it was a small concern, owned by a husband and wife, and the wife was writing to thank me for my business. She seemed genuinely grateful; as far-flung as the reach of the Internet is, it often seems much wider than it is deep. For all the worldwide computer network's vaunted modernity, in human terms there probably isn't much difference between the operator of a modest website anxiously checking the e-mail every few hours to see if anyone has ordered anything, and an old-style small-town merchant gazing out the front window all day long, hoping a customer will wander in. I couldn't imagine this woman's little company was overrun with orders for items such as Toddle House plates.

She had asked me if I wanted a gift card sent with the plate. I took out a pen and tried various versions of notes that I might have her send to Jack—tried to get the words right, to sum up the emotion inherent in all of this. Nothing worked; we were never people to say out loud what we really felt, we liked to hide it behind a joke, and everything sentimental I attempted to write sounded off-key. The Toddle House had never been about sadness. I thought of the five of us sitting on the stools, all those nights and all those smiles.

That's when I began to smile myself, and came up with something that I knew would make Jack do the same. I wrote back to the woman at the antiques store and asked her if she could send Jack a card signed not by me, but by her. She said she would.

Thus, when the package had arrived at Jack's house—the sturdy Toddle House plate safely wrapped inside—this is the note, over the store owner's signature, that Jack read:

> *Dear Mr. Roth—*
>
> *Chuck is not a true friend.*

WE WALKED IN. "CAN I HELP YOU GENTLEMEN?" the young guy behind the register said.

Jack was looking up at the wall, where the menu was posted.

"We're just looking around," I said.

It seemed small, which should not have been a surprise. There had never been any room to wander around the Toddle House—the aisle between the front window and the row of stools had never been spacious enough for milling about, it was just wide enough for you to walk to where you were going to sit. Still, the way the pizza place was configured now made it almost impossible to fathom how much of our lives had once fit into this room. Maybe that's what felt so small about it—it was too small to hold all the memories.

"What would a meal cost, about a dollar ten?" Jack asked me.

That was almost, to the penny, the amount I'd been thinking—$1.10 was the figure in my mind. Cheeseburger, hash browns, Coke. It sounded unlikely, until you broke it down. Fifty cents for the burger, thirty-five cents for the potatoes, a quarter for the Coke. "A dollar ten might be high," I said. "I don't think a Coke cost a quarter."

"Maybe the buck ten included tax," Jack said.

So many times over the years when I'd been in some coat-and-tie big-city restaurant or other—not even restaurants with especially haughty affectations, just one or another out of so many white-tablecloth restaurants in America's most famous cities charging the going rate—and the bill would come, and I'd look at it, and I'd think about this place. Some restaurants you could have a cheeseburger and lyonnaise potatoes (that's what they tended to call hash browns with onions mixed in), and a couple of drinks, and you'd end up paying forty or fifty dollars. Maybe more. Inflation figured into it, of course, and a cocktail costs more than a Coke. But no meal like that, no matter how tasty, ever matched what we'd found in here. And forty years later, here we were.

I looked over at Jack. He was still peering around, studying the wall behind the pizza oven, and the guy on duty must have thought he was continuing to look at the menu.

"We've got menus you can take with you," the guy said, trying to be helpful.

"No, thanks," Jack said. "I'm just trying to figure out if that back wall used to be set deeper."

The guy shrugged. "It's been like this ever since I started working here," he said.

It was as if Jack was trying to take everything with him—he was in search of the ultimate carryout, he wanted to soak in every sight and every recollection, this was a to-go order more valuable than most. "The refrigerator was down there, right?" he said to me.

"Right," I said softly. "It was at the far end of the counter."

"Yeah, the pies had that whipped cream covering on top," he said. "I always loved it when I got the first piece of pie, and I hated getting the last piece."

"The last piece was sort of soggy," I said.

"Not sort of soggy," he said. "A lot soggy."

He stood there, seeing ghosts—ghosts in moccasins, ghosts with their shirtsleeves rolled up—and then, as he always did to show that something had sunk in, he nodded, once.

He looked over at me. "I should get home," he said.

"We are home," I said.

We turned to leave and—not loudly, he didn't want the guy behind the register to think he was making fun—he yodeled the words.

"Toddle House," he yodeled quietly, like a long-ago waitress answering a telephone, like a short-order cook summoning an elk.

WE TOOK THE BACK WAY TO HIS HOUSE, through the alley behind the restaurant again. There was a sign on the brick wall of the parking lot; I hadn't noticed it on the way in, but now we both saw it. SMILE, YOU'RE BEING WATCHED, it said—notification that there was a video security camera.

"Do you see a camera?" Jack asked.

I looked everywhere there might be one.

"No," I said.

"It's probably cheaper just to buy a sign than to actually put a camera in," he said.

Then:

"I'm worried about Janice."

He didn't have to explain. This was taking its toll on her, and it could only be expected to get worse. For all the hopeful phrases and upbeat euphemisms the people around Jack and Janice kept using, the reality of where Janice was headed could be summed up in one cold and unambiguous word, and the word was widow. She knew it and he knew it.

"I wish I could tell her that everything is going to be all right," he said.

He had met her during that period of most people's lives when, no matter how well or how poorly things may be going, they often feel a little unmoored. The first years out of college, there's no road map; everyone and everything you've depended on seem transformed, it's not so much that they've changed as that for the first time they've assumed a different context. You're not where you were anymore, in all kinds of ways; you're en route, at a rate and a pace not necessarily determined by you. The people you've always counted on to let you know how you're doing are no longer consistently present, and even when they are present, they aren't all that sure of how they're doing themselves.

The twenties and thirties are a time, for some people,

when old friends are constantly trying to impress each other with how brilliantly they're faring in business: to verbally one-up each other. It seems, beneath the brash surface, to be partly a product of fear: If you tell someone you're doing great in your new world before the person has a chance to ask, then you're a little less exposed. Maybe.

I don't recall us ever doing that; I don't think the five of us ever fell into that trap. But if we didn't try to snow each other, neither initially were we present in each other's lives the way we once had been. Jack was a schoolteacher in the Chicago area right out of college, and he got married, a marriage that was brief and less than happy. Chuck was doing well in his father's business in Columbus; he met Joyce, he and Joyce introduced Joyce's twin, Janice, to the newly divorced and emotionally raw Jack, and before anyone knew it the two friends were married to twin sisters.

So none of us ever consciously one-upped the others about our successes or lack of successes, but with Chuck and Jack the facts of it were inescapable. Jack was back in Columbus and opened a little bookstore he called My Back Pages, named for the Bob Dylan song. He loved books and he loved the work, loved the independence. Yet Chuck was soaring in the business world, and Jack was living from one week's bookstore receipts to the next, and they were married to twins.

"I know she's really on edge," he said now.

There were always matters that went publicly unsaid about the disparate lives led by the one twin and her husband, and the other twin and her husband. Janice was the love of Jack's life. I had long known there were material things he wished he could give her.

And now he knew that he was going to be gone.

"They didn't need surveillance cameras when we were coming here every night," he said, still trying to find a lens pointed in our direction, still failing to see one.

"I know," I said. "No one was watching us."

"Except us," he said.

SHE WAS WAITING FOR US AT THE FRONT door.

"I thought Greene had put you on a plane to the French Riviera or something," she said. "You've been gone a long time."

"I'm getting a little tired," he said to her. "We walked around a lot."

"Do you feel like eating something?" she said.

"I think I may take a nap for a while," he said.

He climbed the stairs and as soon as he was out of earshot she asked me, "So how did he do?"

I said he'd done just fine, which was the truth, and even as I was saying it I hated the fact that we were all in this new territory where her question was not only pertinent, but necessary. Just a month before, asking me a question like that would never have crossed her mind.

We sat and talked for a few minutes, and I said I thought I'd take off. On the way out of their house I saw where she had mounted the plate.

We had signed it—that night after all of us had met for dinner at the Top, and then had come back to Jack's house, Janice had found a thick-tipped black marker and we had all signed the Toddle House plate. Jack signed it first, and when he had finished the rest of us had signed our names around his.

Our signatures, in that bold black ink, stood out against the white of the plate's china surface. I could see it from across the room—I could see all of our names. Janice had displayed the plate in their home as if it were a piece of priceless artwork, which, in all the ways that suddenly counted, is exactly what it was.

Ten

Allen had gone back up to Canton—there was a trial for which he had to prepare—and before he left Bexley he told me how grateful he was that he'd come down to spend this time with Jack. He—all of us—knew that there was no certainty there would be many more chances like this.

"Jack knows the reality," Allen told me. He had spoken privately with Jack; Jack, he said, was well aware that the odds were not in his favor.

"I told him that we're all going to see each other in five minutes anyway," Allen said. "In the great scheme of things—the infinity of time—it will seem that we all die within five minutes of each other."

"That's a pleasant thought," I said.

"Come on, Bob, it's true," he said. "Jack's sick, but even if he wasn't, all of us are getting there. The game may not be over. But at this stage of our lives, we're all heading up the fourteenth fairway."

WE HAD THE LUXURY OF TRYING TO PUT everything in that kind of perspective—to come up with ways that made all of this seem like part of the abiding narrative of life. And without question there was truth to it.

Jack, however, was waking up every day with his stomach in knots. He had forwarded his office's telephone to his home number, so that any business contacts who called him at work would reach him. Debilitated as he was by the radiation and the chemotherapy, he would set aside time each day to sit at his home computer to work on his sales accounts. He was trying to keep up.

But he knew, without anyone having to say it out loud to him, that the world's patience for other people's woes is fleeting. He was a one-man business operation; he had partners in certain projects, he had longtime trusted colleagues with their own companies, but he was mindful that once a man disappeared from the radar screen of commerce, no matter how understandable the reasons for the disappearance, he ceased to be much of a factor. Occasionally someone might think to say "How's Jack Roth doing?", but people had to make their own livings and take care of their own families, and they would.

"Business is business," he said to me one day, explaining why he felt compelled to take as little time off as he could get away with, to hover over his computer and check his phone messages and go through the mail. He didn't mean it in a cynical way—he didn't say "business is business" the way a harder person might say "dog eat dog." He wasn't talking about his own lust for profits. He meant that if he didn't do everything he could to stay current with his business obligations, someone from some other enterprise would step in to fill the vacuum. Nothing personal, and by the way, how's Jack Roth doing? Business is business.

"I may have car payments to meet, but so does everyone else," he said.

"When did all of that start to become the main thing for everyone?" I asked, already knowing the answer.

"Paying your bills?" he said.

"And what kind of car you have, and what neighborhood you live in, and where you take your vacations," I said.

"The day you start your first job," he said.

Summer was coming. I reminded him of the first taste of the work world we'd ever gotten.

"I guess the first time we worked was that summer when we helped Pongi out," I said.

Jack, of course, had always given his dad a hand at the fruit-and-vegetable market, but he had considered that to be a family obligation. The summer when we had helped our friend Pongi was different. It had seemed like fun, then. It was called work, but for us it was recreation, because nothing was riding on it.

"I liked the feeling of pushing the handcarts into all those grocery stores," Jack said.

Pongi's father had run an RC Cola distributorship in central Ohio, and one summer Pongi had been a fill-in guy on delivery routes. We'd ride along with him—I'm quite certain he didn't pay us; we'd wheel the metal carts stacked with cartons of soda pop into the supermarkets, going from the heat of the day to the air-conditioned chill of the stores, and it was enticing, it was new. That was the best part of it—the newness, or the illusion of newness. New faces in every store, new streets at every turn, a new look in people's eyes, a look we'd never seen before—in those eyes we weren't the kids around our parents' houses, we weren't the same old kids on the same familiar blocks, we were the RC guys, bringing in the pop—if this was work, how could work be so bad? How can you ever get tired of things being new?

"New gets old pretty fast," Jack said.

"If it didn't, we'd still be wheeling RC bottles and still wondering why everyone didn't love to go to work," I said.

"I've got to check the screen," Jack said, as if on it he would find some answer.

JANICE CAME INTO THE ROOM AS WE WERE talking one afternoon and said, "Jack, can I get you a milkshake?"

Her tone was a little too blithesome, and I knew why. He'd been losing weight. He wasn't eating as much as he should. He had no appetite. This couldn't be good.

Her offer, in a June Cleaver voice, was one more instance, although a small instance, of how everything had changed. She knew she had to be careful; if he sensed she was patronizing him it would upset him. He wasn't a child. Yet she had to get him to take more nourishment.

"Maybe I'll have one later," he said.

"Let me make you one now," she said, maintaining the peppy inflection. "I don't think you had any lunch, did you?"

"Fine, then," he said, a little too abruptly. He knew why she was persisting.

She went into the kitchen and he and I pretended that nothing of significance had just happened. She came back with a milkshake—he never would have done it on his own—and she sat with us and the three of us watched a little television and talked about everything and nothing. I could see that she was keeping an eye on the glass she'd brought in.

After about half an hour she said, "Jack, I made you the milkshake and you've taken two sips." Careful even about how she said that—trying to make him believe she was concerned about her wasted effort in the kitchen instead of the truth, which was that she was fearful to her very core that he wasn't eating.

"Jan, I'll drink it when I drink it, all right?" he said, and I hated this, I hated all of our helplessness to change something even this small, this large.

"EVERY FOURTH OF JULY, IT SEEMS THAT I know fewer and fewer people," he said one day.

It was a tradition in town for everyone to gather for the Fourth of July parade, and after that to go to the Jeffrey Mansion—Bexley's public community center and park, bequeathed by the estate of the wealthy civic leader who for many years lived there—for a barbecue, with the mayor doing the grilling. At night were the fireworks.

We always went to the parade together. When we were kids we would sit on the curb at the end of my family's block on Bryden Road. When we were out in the world on our own, and my parents had moved farther east in

Columbus and out of Bexley, I would come in from Chicago for the Fourth and Jack and I would meet at my uncle and aunt's house on Brentwood, where the parade would pass right by their driveway. Now that Jack had built his house on Bexley Park we would watch from the yard of a friend of his on a corner at Roosevelt Avenue.

"We've always thought of it as our town, but I think that's changing," he said. "I used to know every face."

What struck me was not the veracity of what he was saying—it happens to everyone, I suppose; as they grow older and new families move into their communities the faces no longer feel quite as familiar. What struck me was the timing. We weren't at the Fourth of July yet, or anywhere near it. But he was thinking about it.

"You figure in five or ten years, we'll hardly know anyone at the parade at all," he said.

I let that one pass.

"I really wish I knew more of the people now," he said, meaning, I guessed, one of two things. Either he wished he could turn back his clock to a time when his life was not covered with storm clouds and he knew every face and every voice at the parade, at the Jeffrey Mansion, at the fireworks. Or that he wished not that he knew more of the people—but that more of the people knew him. Immortality, and our desire for it, comes in many forms, some accompanied by marching bands on midsummer mornings.

HE TOLD ME ABOUT HIS FATHER—MORE than he had ever said before.

We were talking about the different paths upon which

life had taken us, and I said it had all started when we'd gone off to separate colleges. Up until that time, everything we did, everyone we knew, had played out against the same backdrop. Then I had gone off to Evanston, Illinois, to Northwestern University, where I'd heard they had a program that would teach you to be a newspaperman. Jack had gone to Ohio University, in Athens, in the southern part of the state.

Now, in his living room, I said, "It's funny, I didn't think of the two schools as being any different from each other except for the fact that you went to one and I went to another." Northwestern was a private university with a hefty (for the time—I think it was $600 a quarter) tuition; Ohio University was a state school that wasn't especially expensive.

"The only kind of school that was in the cards for me was a state school," he said. "It's what my dad could do."

We hadn't talked about any of that, when it happened. I have a feeling that Irvin Roth had not spelled out the financial reasons for any restrictions in college-application possibilities for Jack; knowing Mr. Roth, he probably just said one night at the dinner table that Jack should apply to a state school within Ohio, and that was that. No questions necessary or welcome.

"He'd had such a hard life," Jack told me now. "He was born in Poland, and his family was not wealthy, and his father came to America to make a new life as a peddler. But my dad, when he was a boy, did not get to come to America with his parents and his brother and sisters. I think it was because there wasn't enough money."

"So who took care of your dad?" I asked.

"He was the oldest of the children," Jack said. "He had been sent off to live with an aunt and uncle in Poland.

They raised him; he really loved them. He was studying and doing well in school, and he wanted to keep doing that. But when he was fourteen his father sent for him. He didn't want to go. By that time, he considered the aunt and uncle to be his mother and father. He begged not to be sent to the United States."

"And?" I said.

"He had to come anyway. He felt like a stranger to his own family. He arrived in America at the age of fourteen not knowing how to speak a word of English. The whole life he had back in Poland—his school, his friends, everything—he had to leave behind. And he arrived here, with his dad knowing he hadn't wanted to come, and he had to start over."

He had taught himself the new language, and in Ohio, where his father was supporting the family by peddling goods from a cart, Irvin Roth had decided that when he was old enough to get married and start his own family he wanted a better life than that. So as a young man in central Ohio, driven by sheer necessity and self-reliance, he had worked in a small grocery, and then had opened the fruit-and-vegetable market, and he had raised his family, and all those years I had seen him around the house on Ardmore, so silent and so stoic, I'd had no idea. He was just Mr. Roth, Jack's dad, who left for work before the sun came up and who would seem so exhausted at night.

"Did he always regret that he had been made to leave Poland?" I asked.

"He would have died," Jack said. "The Nazis came to round up the aunt and uncle who raised him. They were exterminated in a concentration camp. If he would have stayed, it would have happened to him."

His father had said a state school in Ohio would be the

place for Jack. He offered no reasons, but then, Mr. Roth always kept most things to himself.

I THOUGHT I'D HEARD ON THE RADIO THAT the Clippers were in town for a home stand, and I asked Jack if he'd like to go to a game tonight.

"Maybe," he said. "Let's see if I have the energy."

The Clippers were Columbus's minor-league baseball team. They played in an old ballpark on West Mound Street, in a neighborhood that had seen vastly more well-to-do years. They were born the Columbus Red Birds, became the Columbus Jets in the jet-age days when we were boys, and now were the Clippers. The one thing that had remained constant was the cemetery just past the left-field fence.

"Bring your Jet Badge," I told him.

He smiled. When we were in grade school you could purchase a Jet Badge—round blue metal badge, metal pin on the back—for fifty cents. On special nights throughout the season—doubtless they were evenings when attendance was expected to be dismal—your Jet Badge would get you in free, if you had a paying adult with you. It wasn't an unclever piece of marketing—the Jet Badges (or, more accurately, the kids on whose shirts the Jet Badges were pinned) were the lure to get the dads to buy tickets on nights those dads would otherwise be sitting at home staring at the television. The kids with the Jet Badges couldn't get in without an adult ticket bearer; the dads may have been there under some duress, but they were there. And the baseball wasn't bad.

At least we didn't think so. "You'd have thought we

were in Yankee Stadium the first few times we walked into Jet Stadium," I said to Jack.

"What did we know?" he said. "The players were in clean uniforms and they could hit the ball."

Over the years we'd go back many times, starting as children and stretching right up until now. Our evolving view of what the ballplayers represented fairly accurately mirrored our view of what life might have in store for all of us.

"When we first went to Jet Stadium I thought the players on the Jets had made it to the very top," Jack said. "Getting paid to play baseball in front of fans."

"It didn't seem that way when we were in our twenties and thirties," I said.

By that point in our lives, we'd go to Jet Stadium—it had been renamed Cooper Stadium by then, in honor of Harold Cooper, a former Jets general manager who had become a county commissioner—mainly as a lark. We'd sit and talk and drink beer, and the games weren't really the point.

"I think we saw the players for what they were by then," Jack said. "Guys who were striving for something that didn't have much chance of happening."

That was the reality of the minors. Some of the players would make it to the big leagues, most wouldn't, and if you were ever able to step onto the field and look at them up close, you'd probably see the frustration and fear behind some of their eyes. Especially the older ones—there is a poignancy to an old minor leaguer who has never been called up, who keeps waiting while the newer, quicker, stronger players move past him. . . .

"I always sat there and thought about the ones on the way down," Jack said.

There weren't all that many of them—players who had

made it to the majors for a short time or even for a long time, who had been released and told they were no longer wanted, but who persisted in playing minor-league ball. Those were the ones, when we were in our forties and sitting in the stands on Mound Street, whom we'd talk about. Not so much the players who had been sent back down for physical rehab, or to try to sharpen their swings and make it back to the big leagues. In our forties we would look at the men who were there by choice—the older players who had been to the majors and knew they were never going back, but who elected to keep playing baseball in the minors anyway. Either because they loved the game, or because they simply couldn't conceive of anything else to do.

"At some point," Jack said, "you start to figure out that what you're seeing on the field is guys who really want something and who probably are never going to get it."

Not so different from most lines of work, except people pay to watch, and they eat peanuts while you succeed or fail. Quiet desperation, with the cemetery beyond the outfield fence.

THIS TALK OF THE FOURTH OF JULY, OF the Clippers and the old Jet Stadium . . . summer wasn't here, but summer always meant hope, summer meant good things straight ahead. Some of Jack's words may have been downcast, yet the fact that he was talking about summer—that it was on his mind—meant something. If you're thinking about summer, there's the best kind of anticipation in your heart.

So I was sort of blindsided when, out of nowhere, he said, "Greene, do you think it could be the DDT?"

We had joked about it for years. The DDT trucks had been an indigenous part of our summers. They would rumble down the streets of Bexley, on their appointed rounds, spewing out thick white clouds of DDT insecticide in the years before DDT was banished in the United States. Not only would our parents not keep us away from the DDT trucks, we would run after those trucks, running right into the DDT smoke, smelling it, drinking it, sucking it in. It was delicious, it was the fragrance of summer, we thought that the smell of DDT was the smell of fun. We ate it. Our mothers and fathers warned us not to run too close to the back wheels of the DDT trucks, in case the trucks were to stop suddenly and back up; our parents didn't want us to get run over. But the DDT itself, the DDT that was billowing out of the rear of the trucks' big tanks—that was swell, that was OK with everyone, summer meant ice cream and DDT.

That's what we had joked about in later years, in our thirties and forties. Can you imagine that? we'd say to one another. We drank down the DDT clouds, day after day. No one knew we shouldn't.

Now Jack was asking me, and it wasn't that his voice was plaintive—it had a tone more of curiosity—but he was asking on the straight. It was no joke.

"I keep thinking, what could it be?" he said. "I never smoked. I always ate right. Where did I get so sick?"

Summer was on its way. Season of hope.

HE WASN'T UP TO GOING TO THE BALL-game. He told me at the last minute he just didn't feel well enough, and I said not to worry, to get some rest. I didn't know what I was going to do with the night, and I decided to call Dan Dick.

It was an unexpectedly awkward invitation to extend—something that forty years before would have been second nature now caused me to hesitate. This was a Saturday evening; Dan and his wife lived out near Buckeye Lake, and I didn't want to intrude on their plans by calling him at the last minute.

A weekend night together when we all first were best friends was a given. That's what the Toddle House was for. But—save for our ABCDJ dinner the other night—I hadn't seen Dan much in recent years, and I felt vaguely clumsy ringing his phone, once the easiest thing in the world.

His wife, it turned out, had been away from the city for the day, and Dan had nothing going tonight. I asked if he wanted to do something, and he started to recommend some restaurants, naming different types of food, as if the eating was the point. I said, "We don't have to plan a place, Dan. Just pick me up."

Even that felt clunky as we talked about the details—we hadn't hung out on a Saturday night in so long. "Do I come to your room?" he asked. "Just pull up to the front of the hotel," I said.

And so Dan and I were soon driving around, no destination in mind, as we had so many times before. He made me laugh right away, even though that was not his intention; I had told him Allen Schulman's comment about all of our encroaching mortality—I thought Allen had been right on the money, and in a way that I hadn't considered before. Jack might be sick, but all of us, at this stage in our lives, were trodding along the same inexorable path.

"So what Allen said about all of this," I told Dan, "is that we're all heading up the fourteenth fairway."

Dan, behind the wheel, nodded.

Several seconds passed.

Then he said:

"I don't get it."

"Don't get what?" I asked him.

"The fourteenth fairway," he said.

I explained to him what Allen had meant. Life is a journey, and we're all on it, I said. It has its seasons, and once you get older you know that the end is on its way. The course has only eighteen holes, and although you can't predict anything with certainty, once you're deep into your fifties you and your friends are all pretty much at the same place on that course. You're on the back nine of life, you're heading up the fourteenth fairway.

"So you see?" I said to him.

"Well, I guess," he said. "If you're a golfer."

I laughed out loud and he turned his head and looked at me and he started laughing, too; we drove with no purpose and it occurred to me there was something he might not have been aware of. I mentioned it to him, and he didn't in fact know a thing about it, so we drove there and got out of the car and stood together looking down at it. The ABCDJ brick, the one Jack had bought to honor our friendship. "He's something," Dan said. "He's something."

WE GOT HUNGRY AFTER A WHILE, AND I said I'd heard there was a Smith & Wollensky steakhouse out at the sprawling new Easton shopping development.

Dan wasn't quite sure how to get to the restaurant; he became lost and ended up calling out the window of his car to prom couples on their way to parties. ("You two look great! Do you know how to get to Smith and Moresko?") I'd known Dan long enough to realize it would do no good to try to correct him—Smith and Moresko was as close as he was going to get to Smith & Wollensky this night, I was laughing so hard I couldn't see, and soon enough we were in a little strip mall that wasn't even part of the Easton complex, and Dan kept calling out the window to strangers, searching for the elusive Smith and Moresko, and the sound of the laughter in the car was like music, like a long-lost song.

We finally found the place and waited at the crowded bar for a table. There were attractive women all over the barroom, and Dan, with great seriousness, said, "They're all from *The Swan*," and without asking what he meant I knew what he meant. He was referring to that television program that offered cosmetic surgery to female contestants, and he was trying to say that it's possible for women to make themselves look any way they want these days—it wasn't that these particular women in the steakhouse had had cosmetic surgery, and certainly not that any of them had actually been on the television program. Only one person in a million would have had any idea what Dan was talking about, but I was that person; in the context of Dan's lifelong lexicon, starting with Reedeep Reeves, what he'd said made perfect sense to me, and I told him that I agreed, they were all from *The Swan*, and we were called to our table.

It was like exhaling, the loose and unstrained time together, and when we departed there was some small confusion about the bill—we'd left the money on the table, but apparently the waiter hadn't seen it. As we got into Dan's car outside a manager came running out of the restaurant to find out what had happened and to try to clear things up, and now I was really laughing, my eyes stung, and Dan gunned the car and drove away and asked me what was wrong. Nothing, I said—nothing. It's just that the manager running out to cut us off, the mix-up about the bill—it's forty years later, I told him, and we're still being accused of stealing the chicken from the Latin Club banquet. I was laughing and trying, through the laughter, to get the words out—it's the chicken at the banquet, I said, we're fifty-seven years old and it's the chicken at the banquet.

"I had a filet," Dan said, which is the only thing he could have said, which started me up all over again. He dropped me at the hotel and I walked to my room and opened the door and there was dead silence, no voices. Now I thought about Jack again. The time with Dan had been temporary respite, but I stood in the doorway of the room and there were the beds, neatly made up, there was my suitcase, and I thought about my reason for being here; the laughter was done for the night, and I've seldom felt so solitary, so alone.

Eleven

I WENT BACK TO CHICAGO, AND THE FIRST thing I did when I walked into the house from the airport was to call him. It seemed essential to check to see if he was all right, to hear his voice—essential in a way it never had been before. The next day he called me, and I called him later that evening and again the morning after that. Without either of us saying it we both seemed to know that we had decided something, although decided is too formal a word, it's a word that denotes conscious thought, and this was something more elemental than that. But even without saying it we both knew what was going to happen, which was that we were going to talk every day for the rest of his life.

"I'M WORKING ON SOME NEW LABELS FOR this olive oil that I'm naming after Maren," he said. "They really look sharp."

The commercial particulars of the olive oil named for his daughter, I knew I wouldn't be able to figure out. That's the kind of thing he did for a living—devise ways to take basic products, products most people don't even think much about, buy them up in large quantities, package them imaginatively, get them distributed to major retailers . . . the kinds of business intricacies that are second nature to some people, but have always, to me, been as mysterious as quantum physics. How Jack had taught himself to know how to do that kind of thing, and had made a living at it, was a great puzzlement to me. One day we're tossing socks at those shoe-box baskets above his bedroom door, the next he's custom-designing labels for privately bottled olive oil.

But it was that one phrase coming from his voice over the telephone line—"They really look sharp"—that made me grin. Not so much the concept that a label on a glass container of olive oil could look sharp—if Jack said his label was sharp, I had no doubt that it was sharp. It was the word itself—*sharp*—that delighted me as he said it. You seldom hear that word in conversation anymore. He'd been saying it for fifty years.

He'd always used it as a synonym for "cool," or "good"; from the time we were boys, his way of expressing approval for something was that one, brisk word. Back then, he might use it to describe a madras shirt; we'd be shopping for school clothes down at Lazarus, and he'd pick a shirt out of a pile, he'd look at it for a long moment, then pronounce: "Sharp." It was always a part of him, like that chipped tooth. I couldn't hear it without thinking of him.

So I was happy to hear, over the phone, that he was enthusiastic about something he was working on—any enthusiasm in his voice was very good news, if he was excited about something, if he was invigorated, that meant he had been able to push into the background, at least for one part of one day, some of his apprehensions. I was even happy that his enthusiasm was about an olive-oil label—whatever worked was more than all right with me. Mostly, though, I was happy to hear his voice saying "sharp." If I'd somehow been able to put that sound in a scrapbook, I would have.

Not that I told him about it—not that I even mentioned to him that I had noticed him saying the word. He didn't know the word was a special part of who he was; we had never discussed it. If I'd pointed it out to him now, he'd probably get self-conscious and never say it again.

WE GREW UP ON TELEPHONES WHEN TELEphones were a pretty big deal. In that "get-off-the-phone" era of dads banging on the bedroom door when you'd been tying up the house's only line, in those five-minute-limit-on-the-phone days when parents would scrupulously apportion out when and for how long the younger members of the family were allowed to talk, telephones possessed an aura that has been transformed, if not completely diminished, in the new age of cell-phones-for-everyone, text-messages-throughout-the-day-and-night, communication-as-a-birthright technology.

So when Jack and I first started to call each other, the power of a phone call was considerable. Mostly, it was because we were so young; when you're five, and one of your parents answers the ringing phone and then tells you it's for you, you can barely believe it. You're almost afraid to touch the receiver. It's for *you*? What are you supposed to do?

You get used to it soon enough; you stop being startled that anyone would think to call you. Three memories of Jack on the telephone, each of them having to do with girls:

In fourth grade, he had a crush on a girl named Sharon Weiner. He had a bunk bed in his room, and he sat on the bottom bunk, the family's black upstairs phone pulled into his room with its long cord underneath the bottom of his door. He had shut that door and locked it, so he could have privacy while he talked (privacy from his family, that is; I was in the room, I was permitted to be there, the concept of privacy did not extend to me). He dialed her number, she answered on the other end, and this is what he said to her:

"Hi, honey."

It was staggering. I could scarcely process it. "Hi, *honey*?" He might as well have been in a romance-in-Monte-Carlo movie starring opposite Anita Ekberg; he might as well, at age nine, have been wearing a satin smoking jacket and feeding bonbons to a platinum blond chorus-line cutie. "Hi, *honey*?" He'd sat on the bottom bunk calling her "honey," I'd sat on the top bunk with my mind in a frenzy, and he had *called her honey*—he had actually, on a genuine, working telephone, been brave enough to dial the number of an actual living girl and call her "honey." Sharon Weiner undoubtedly did not know that there was an audience for this momentous event, but there was: me. Sharon Weiner. Honey.

And:

Years later, when he and I were sitting in his living room, each having no luck in calling girls with whom we wanted to go out that weekend—the girls with whom we wanted dates had informed us without any apparent deep regret that they had previous engagements—Jack's sister Helen (she of Chuck Berry and drop-the-coin-right-into-the-slot) came down the stairs, looked at us sitting there with expressions of glum dejection on our faces, and, seeing us, seeing the phone, deducing what had happened without having to ask us, said in a bright yet gently taunting tone:

"What's the matter, guys, got the girlie hang-ups?"

The perfect phrase, somewhere between sadness and resigned acceptance—the girlie hang-ups is what we had, all right, and in more ways than one.

And:

At some point during that same epoch—the time of our young lives when we strove to do everything we could to avoid the moment when a girl might, in fact, literally or figuratively hang up on us—we devised a way to mitigate the potential sting. If there were girls each of us wanted to go out with, we'd give the names to each other. Then Jack would call the girl I hoped to date, say he was me, and ask her out; I would call the girl he hoped to date, say I was Jack, and ask her out. The possibility of a girlie hang-up was not eliminated, or even lessened. But it didn't feel quite so personal. If she said no, she was saying no to the other guy—the one we'd said we were.

Now we spoke on the phone each day, often more than once, and our calls had been so customary for so many years that most of the time it was difficult to remember when this was so new to us and when the very fact of a phone call was laden with drama. The unspoken drama behind our daily phone calls now, necessarily, was of a different kind.

CHUCK HAD LEARNED HOW TO USE THE instant-message function on his computer, and also how to be aware of when I was signed on to my own computer at home (if I'd known how to block this information from being sent out, I would have), and he had taken to annoying me whenever he saw I was logged on.

This was all a part of his running who's-really-Jack's-best-friend gag, and he upped the ante once I got back to Chicago.

I'd be working on my computer and the instant-message box would pop up and Chuck would write: "Jack doesn't want to see you."

I'd click the little thing to make his IM go away, and then another one would pop up: "He said you specifically."

I'd get rid of that one and immediately he would send a new message: "Why did you come here?"

We couldn't see each other's faces; because of the computer communication, we couldn't even hear each other's voices. But the joking, over the miles, was serving the same purpose that our jokes had always served, from the summer nights in Allen's blue Ford right on up to the present: to hide our feelings—to pretend to feel nothing, when pretending to feel nothing seemed much better than the alternative.

So Chuck would do this all the time now, and what I knew, what we never talked about, was just how much pain Jack's own pain was causing him, just how much all of this was tearing him up inside. He was still Chuck; he was still the guy who wanted the world to think there wasn't a thing that could ever get to him. One morning we'd been instant-messaging back and forth about nothing in particular—there'd been no jabs from him this day, just beginning-of-a-new-morning how-are-you-doings—and then he sent the message:

"I've got to go now. I've got to drive Jack to get his head zapped."

That's what Chuck would do these days. Drive Jack downtown to the hospital for his radiation or his chemotherapy, take him in, wait for him, drive him home. And that's the way he would tell me about it. As if it was just a nonchalant and easygoing part of his routine, instead of an almost sacred duty he had undertaken. Which it was, and which he was carrying out with a depth of devotion that sometimes moved me in ways that made me bow my head in admiration.

THE SUMMER I HAD MY FIRST NEWSPAPER job—as a copyboy at the *Columbus Citizen-Journal*, between my junior and senior years of high school—I worked nights, and one of my responsibilities was to put together the emergency runs and the fire runs each evening.

This was basic work: Call all the local and surrounding fire departments and police departments, ask them when and for what reasons the ambulances and fire trucks had gone out, type it all up and give it to the copydesk. Time of day, address, names of people who had needed help—nuts and bolts, and information that today likely would not be given out so blithely, in the new and rightly circumspect era of privacy laws.

I'd make the phone calls and talk to the cops and dispatchers, I'd type it up, and it would be in the next morning's paper, in little agate print.

Jack wanted to be in the paper. He told me he *really* wanted to be in the paper.

So one night—in an act that now, with good and undebatable reason, would almost certainly get the person who did such a thing burned alive at the stake in the Journalistic Ethics Town Square—I did it. Egged on by Jack, I typed up the following item, in the middle of all the day's other emergency runs:

2:17 p.m., 228 S. Ardmore Road, Bexley,
Jack Roth, 17, cut head. Treated and released.

It ran the next morning. I felt terrible almost as soon as I did it; what had seemed like a harmless prank struck me, by daybreak after a sleepless night, as an awful misjudgment—right up there with stealing the bottle of 151-proof rum, but this time with the power of the *Citizen-Journal*'s printing presses flinging proof of my transgression far and wide, onto doorsteps all over central Ohio. I waited for my moment of reckoning, which I assumed would come from a none-too-happy Bill Moore, the city editor for whom I ran off to Gray's Drugs at Broad and High each day to buy his thirty-five-cent packs of Lark cigarettes.

It never happened. An aunt of Jack's read the emergency listing about him and became so upset she nearly swooned, but other than that there was no notice taken of it, and thus no retribution. It had been a joke—Jack had thought it would be funny to say that his head was hurt. It had been one of the big jokes of that summer.

"SO I WENT TO THIS PICNIC TODAY THAT I always go to," he said.

I could tell something was wrong. I could hear it in his voice before that first sentence was finished.

He said it was a picnic put together by the same community group every year. He always pitched in at the annual picnic, he said; it was something he enjoyed doing.

"I went and I cooked hot dogs for everyone," he said. "That's usually my job there."

"How did it go?" I said. So far, the details of the story did not match the fragile sound of his voice over the phone line.

"The picnic was fine," he said, "and then a group of people were going to get together at Mark Masser's house afterward. Janice was driving me there.

"And I just started crying. . . ."

His voice trailed off, and I could tell that he was doing his best not to cry now.

"It just started to happen . . . ," he said, and he lost his effort to hold it back, and I sat and waited.

"It's happened a couple of times before, and I don't even feel it coming," he said. "And I just can't stop."

"Do you think it was because of anything in particular?" I said.

"I see those people every year, and I look forward to it, and I was thinking about . . ."

He couldn't go on.

I waited and then I said: "Next year?"

I knew that, hundreds of miles away, he was nodding his head.

A few more seconds passed and then he was able to speak again.

"I thought about really wanting to see those people next year, and I may not," he said.

We sat there on the phone line, each of us holding on.

THE DAY HE FIRST SAW OHIO UNIVERSITY—after his dad had told him that a state school was where he would be going, after he had applied and been accepted, after he had received a mailing telling him in which dormitory he would be living—he took me with him.

This was during the summer before the fall term began; incoming freshmen were invited to come down for a day of orientation, and to sleep in their dorm for a night. His father had bought him a car—a little used green sports car, it was either a Triumph or an MGB. It was very unlike Irvin Roth to let Jack have something that racy, but I perceived there was a feeling of letting go—Jack would soon be off at college, on his own—and his dad had said yes. We'd all stood in the crushed-stone lot of a used-car dealership on Main Street, and I remember Jack's dad asking the salesman if the car had a roll bar, and being silently surprised that Mr. Roth would even know to inquire about such a thing; "roll bars" were the stuff of drag strips and car-racing magazines, but apparently Mr. Roth had read up, and wanted to make certain Jack was safe.

So we were on our way to Ohio U. in that little sports car, zipping south on Route 33, and there was a new Beatles song out that week: "I'm Down," with McCartney singing a high, almost falsetto lead. We talked and we looked at the map to be sure we were on the right path to Athens, which really wasn't necessary, it was a straight shot on 33. McCartney sang the song we were hearing for the first time—this was during that period when new Beatles singles were coming out so frequently that it was hard to keep them all straight—and his words pulsed insistently out of the car radio: . . . *you don't cry 'cause you're laughing at me* . . . We passed small towns and family farms and wooded hamlets, and even in the car, even before we got to Jack's school, we sensed that everything was changing.

The dorm, when we arrived there, was mostly empty. Jack found the room to which he'd been assigned; there was a bunk bed with bare mattresses, and we threw the small suitcases we'd brought onto the beds and went out to explore the campus. I can't recall much about the day and a half we were there, except there was a dark little bar in town, called the Starlight or Starlighter or something like that, and they sold us beer without asking us for IDs, which should have made me feel a measure of excitement but did not; Jack was checking out his college and I was just along for the ride, and my mood was a little gray.

We slept on the mattresses that night—we'd forgotten to bring sheets or blankets—and the next day, registration materials packed into his suitcase, Jack climbed behind the wheel of the sports car, I got into the passenger seat, and we drove north on 33, the same road we'd taken down the day before, but something felt different. We were together for now, but we knew it was ending, at any rate it seemed to be. We didn't know what was to become of our old life and of our friendship; he was going one place and I was going another, and the radio was on all the way back to Bexley and I don't think we said very much at all.

"I HAVE A REALLY GOOD VIEW."

He was telling me what he saw during chemotherapy. He was actually trying to make me think there was something pleasant about it.

"I sit in the same chair every time, and it's a pretty comfortable chair," he said. "It's right in front of a big window that looks out upon downtown, and there are some trees on the street . . . I like the view. It's sort of relaxing."

He would fill me in on the logistics of the chemo—how the chemicals that were dripping into his arm would be calibrated to enter his body at a precise rate over the course of a number of hours; how there was music available, and television screens within his sight line; how it didn't hurt, and the time passed more swiftly than he would have anticipated.

He told me the bathroom routine, and what some of the other patients receiving chemotherapy were like, and each time we talked about it he would return to describing the view: "I like looking out at the trees," he would say.

"Does it get kind of boring?" I would ask.

"Well, it gives me time to think about some things," he would say. "So I feel I'm accomplishing something while I'm there. I wouldn't say I get real bored."

Janice or Chuck would always be waiting when he was finished, to take him home, he said. "That's when it starts to kick in—I get real tired and kind of sick once I'm home."

But the view, he said, was pleasing to him: "Considering what somewhere like that could be like, I've got a really nice place to sit."

I don't know if he believed it; I have a feeling he was saying it in the hopes that I wouldn't feel so bad about what he was going through.

"You can watch DVDs," he said, the one person I have ever known who would try mightily to find good even in this.

AND THERE IS THIS:

It may not mean all that much, but for more than fifty years, whenever I would hear his voice on the phone, or he would hear mine—just that first "Hi," that first "What's going on?", that first "Hey" . . .

Whenever that would happen, when the phone would ring and one of us would pick it up and hear the sound of the other's voice, we never had to say who we were, or ask. Not once.

It happens, among the best of friends; hundreds, even thousands, of people come in and out of your worlds, business associates and colleagues change, you move from one city to another and become a part of new work and social circles. Your ears are filled with a cacophony of voices, so many voices overlapping and competing for a person's attention over a lifetime.

Thousands of them. Yet it can happen:

After half a century, a voice across the miles can say "Hi." And you never have to ask.

You know.

Always.

Twelve

One call I received from him, in the days right before I went back to Columbus, carried a significance I did not immediately comprehend.

I must have been out when he called, because he left a voice-mail message. Just two sentences:

“Greene, I’ve got a question for you. Call me.”

I thought it might be a trivial one—the kind of question with which we had always peppered each other. Questions about what one-hit singer sang what obscure song (Bobby Jameson, “I’m Lonely,” which for a few brief months had been an obsessive, play-once-an-hour favorite of disc jockey Jerry G. on Cleveland’s strong-signal KYW) or what actor had starred in what only-on-for-one-season television series (Nick Adams, *Saints and Sinners*, about the newspaper business). Questions we meant to entertain each other.

This was different. I should have known. "Greene, I've got a question for you," and there was a sense of urgency.

THERE ARE A HANDFUL OF PEOPLE, DURing your lifetime, who know you well enough to understand when the right thing to say is to say nothing at all. When the right thing to do is just sit there with you—either in the room, or on the other end of a telephone line. To be there.

Those people—and regardless of how lucky you are in your friendships, there will be, at most, only a few of them during your life—will be with you during your very worst times. When you think you cannot bear that with which the world has hit you, the silent presence of those friends will be all you have, and all that matters.

When, during an already painful juncture in my life, my wife died, I was so numb that I felt dead myself. In the hours after her death, as our children and I tried in vain to figure out what to do next, how to get from hour to hour, the phone must have been ringing, but I have no recollection of it.

The next morning—one of those mornings when you awaken, blink to start the day, and then, a dispiriting second later, realize anew what has just happened, and feel the boulder press you against the earth with such weight that you truly fear you will never be able to get up—the phone rang and it was Jack.

I didn't want to hear any voice—even his voice. I just wanted to cover myself with darkness.

I knew he would be asking if there was anything he

could do. But I should have known that he had already done it.

"I'm in Chicago," he said.

I misunderstood him; I thought he was offering to come to Chicago, and I was going to thank him but tell him that wasn't necessary.

That wasn't it.

"I took the first flight this morning," he said. He had heard; without calling, without being asked, he had made a plane reservation to Chicago and had flown in. He was already here.

"I know you probably don't want to see anyone," he said. "That's all right. I've checked into a hotel, and I'll just sit in the room in case you need me to do anything. I can do whatever you want, or I can do nothing. I'm here as long as you need me."

He meant it. He knew the best thing he could do for me was just to be present in the same town, to be ready in case I wanted to ask for his assistance. To tell me not to feel obliged, not to give it a thought. He would be there, he said. He wasn't going anywhere.

And he did sit there—I assume he watched television, or did some work, but he waited in that hotel room until I gathered the strength to say I needed him. In those first days, he helped me with the things no man ever wants to need help with—how much do you have to trust another person to ask him to carry your wife's death certificate to where it must be taken?—and mostly he sat with me and knew that I did not require conversation, did not welcome chatter, did not need anything beyond the knowledge that he was there. He brought food for my children and by sharing my silence he got me through those days.

So I should have known, when I received his voice message now—"Greene, I've got a question for you. Call me"—that it might be important. "Call me." That was what should have told me.

"DO YOU THINK YOUR MOM KNOWS ANYthing about how to get long-term nursing care?"

That is what he wanted to ask me; that is why he had made the call.

It was basic, and it was something he was going to have to deal with, and he didn't know how. He was going to get sicker—of that, he was sure; the time was going to come when he and Janice could not take care of all his medical exigencies at home themselves. And he didn't know quite where to start.

He had no illusions that I would know—I'm obtuse about even the most elementary things like that. I don't know how to fill out an insurance form. Long-term nursing care is not a need you consider until that need is right upon you. So he asked if my mother might be able to help him.

How difficult it must have been for him to say those words to me. After a lifetime of talking with each other, conversations that once had been about soda pop and sports teams, about movies and milkshakes, how difficult it must have been for him to ask me about securing nursing care for the time when he could no longer care for himself.

He didn't know what to do. I didn't know what to do.

But he knew that I knew someone who always knew everything.

MY MOTHER IS THE MOST CAPABLE PERson I have ever met.

Had she decided to do something other than dedicate her life to being a wife to my dad and a mom to my brother, my sister and me, I have no doubt that she would have risen as high in the business world as she might have desired. She was a woman of the era during which the standard choice for wives, if they and their husbands could afford it, was to stay at home and raise the children; that she did, with love and great wisdom. She also became the head of numerous civic organizations on a volunteer basis, leading them with the skill of a seasoned CEO. She is a beautiful writer; although for much of her life she wrote only for her family, once she passed her eightieth birthday and was, for the first time in almost sixty years, alone, a widow, she wrote a book that made it onto the *New York Times* extended bestseller list, and followed it with another. There is nothing, I have long believed, that she cannot do.

For Jack, on all the occasions after the age of fifteen when he would have loved to share his happiness, his grief or his hardships with his mother, there was no one there. Graduation from high school, graduation from college, honors, achievements, wedding day, birth of a child—whatever instincts the rest of us who were his friends had at moments like those, instincts to turn to our mothers, Jack had learned to bury those instincts inside. All the times during our twenties and thirties when the rest of us picked up the phone to call our mothers when we had good news about business, or sorrowful news about setbacks, when we just wanted to hear our mothers' voices to reassure ourselves that on the solid and steady level that counted most, they were still there . . . Jack was unable to give himself that. When he wanted to hear his mother's voice, to ask for her advice, he knew he could not. He was, in that most consequential precinct of the human experience, a man alone.

His sweetness of spirit—his core of kindness—at times obscured something else in him: an inner-driven confidence in himself that had been born not of unalloyed high self-regard, but of necessity. Once his mother was gone, and, later, his father, Jack knew that he was all he had—if he couldn't figure life out, there was no one to turn to. He believed he would always in the end prevail because he had no choice but to believe that. The only voice cheering him on was the voice he heard within himself. The first voices—the voices of a mother and a father, the voices that most matter—were both gone.

So when he asked if I thought my mom knew anything about how a person might secure long-term nursing care, it signified, to me, something pervadingly affecting. He was turning to my mother because he had seen her for fifty years, he knew just how smart and how determined and how relentless she could be when the need to accomplish something was vital to her, when that need mattered to her. And he knew that he mattered to her—that he had mattered to her since he was five. He knew that if she didn't know anything about the logistics and regulations of nursing care, the prudent steps to take and the minefields to avoid, she would not rest until she found out. He knew there was no one he would rather depend on.

IT DIDN'T EVEN TAKE ME ONE SENTENCE.

I wasn't twelve words into explaining that he had called me to ask if she could help before she interrupted me.

"I know exactly who to call," she said. "First I want you to ask Jack and Janice some questions about what kind of insurance they have."

She was eighty-five years old. She sounded like a woman on her first day at an important job out of college—in her voice, she had that kind of resolve, that kind of poised volition to succeed.

I told her I would pass the questions on. She said she had some phone calls to make. She had some ideas. She wanted to get started. She said to tell Jack not to worry.

ALL OF THIS WAS GOING ON IN THE TWENTY-first century, and Jack and I had always assumed we would have everything figured out by then.

We were born in 1947, and for years when we were boys we would occasionally have long discussions about what we would be like at the turn of the century—what kinds of jobs we would have, who we would be married to, where we would live, who our friends would be. We knew one thing: We knew we'd be fifty-three (fifty-two when 1999 turned into 2000; fifty-three by the end of March that first year of the new century).

It was always a toss-up as to which was more unfathomable: the concept of living in the twenty-first century, or the concept of being fifty-three. Probably fifty-three; we knew approximately what the basics of our lives would be in the year 2000: robots to clean our houses, personal hovercrafts to take us to work, freeze-dried steak-and-potato pills to pop into our mouths and wash down instead of eating meals, vacations on Mars. We read *Popular Science;* we understood what American life was going to look like.

Fifty-three, though, was a more difficult notion to imagine. Only one thing about that seemed certain: We would know pretty much everything. By fifty-three, we would have the answers. Once you'd lived that long, there wouldn't be anything left to figure out.

When we were ten or eleven we would ride the Broad Street bus from Bexley to downtown Columbus, and usually at some point during our day we'd walk one block east past the Ohio Statehouse until we got to the corner of Broad and Third Streets. We'd look up at the big electrical sign on top of the headquarters of the *Columbus Dispatch;* in orangish red Old English neon lettering the sign announced: "Ohio's Greatest Home Newspaper." To us, that building seemed to be the font of all worldly knowledge and fact-infused mightiness; we knew just about nothing about anything, we were two kids standing on a street corner and peering up at a bright sign, and directly beneath that sign, inside the building, the people at work surely knew everything.

Just as one day we believed we would; by the time the century turned, by the time we were fifty-three, we'd have it all down pat. Of course when I did make it into that newspaper building seven years later, for the summer copyboy job I loved so much, it was to work not for Ohio's Greatest Home Newspaper, but for the *Citizen-Journal,* a smaller and less prosperous daily that shared the *Dispatch*'s office space and printing presses because of a joint-operating agreement worked out with the federal government to prevent newspapers like the *Citizen-Journal* from keeling over and dying (which, in fact, it did do in 1985, once the joint-operating agreement expired); when I did make it into that newspaper building, the repository of all the sagacity and deadline-driven information that our vast planet could provide, what did I do? Put Jack's name in the emergency runs for a cut head that was never cut.

And when the year 2000 finally did arrive—accompanied not by steak pills or hovercraft, but by jitters that everyone's personal computers (personal computers! Why didn't *Popular Science* tell us about *that*?) would break down at midnight because of improper coding, jitters that turned out to be without merit. . . .

When the year 2000 did arrive, Jack and I, and probably everyone else in the world, discovered what we should have guessed all along—that in the new century we didn't know everything after all. We still felt as if we knew, if not nothing, then not nearly enough. Just when we got to the point at which we had thought we would have all the answers, mostly we had questions. More questions than before.

NEW INVENTIONS, THOUGH, WERE SOMEthing he instinctively understood and embraced. I was always slow to accept changes in the machines and products that infuse our daily lives; it didn't seem to me that they would work, never mind be an improvement. Jack, in our thirties and forties, was invariably among the first to try the new way—and would invariably do his best to bring me around.

"Where does the sound come from?" I asked him when he purchased one of the earliest CD players and showed me a compact disc. He explained that the mystery of how music emanated from a CD was no more or less mystical than whatever marvelment made sounds come out of a vinyl record. This was the future, he said.

"It can do everything a record album can, and it sounds better," he said.

"How do you go from one song to another?" I asked.

"Trust me, it's easy," he said.

"But what if you want to start a song in the middle?" I said. "On an album, you can put the needle into the grooves halfway into a song."

"First of all, how many times do you actually do that?" he said. "And second, if you want to with a CD, you can. Trust me. You'll get used to this."

Same with cash machines. He'd needed some money one day after the bank had closed, and we drove over to a branch on Main Street. He inserted a card into one of the first-generation ATMs, and I said, "What are you doing?"

He showed me how it worked. To me, a bank was supposed to be what it had been when we first walked into one: marble floors, hushed tones, a discreet teller writing figures into your passbook. "How do you know they took the right amount of money out of your account?" I said after his cash had spit out of the ATM's metal mouth. "What if you took out fifty bucks and they say you took out five hundred?"

"They don't," he said. "This is just as accurate and much more convenient, and you can do it twenty-four hours a day."

"But what if they say you took all your money out, and you know you didn't?" I said.

"These are the same things you worried about when

they first had drive-through teller windows," he said. "You'll get used to this. Trust me."

We did this mostly for effect; I wasn't quite as nervous about the new ways as I pretended to be, but I was a lot more skeptical than he was, and he seemed to get a kick out of showing me the ropes. Whether it was VCRs or e-mail or cell phones, he was always months or years ahead of me in signing up, and he would always laugh at my reluctance to jump into the pool.

He would travel on business, and he would tell me about the great deals he would get using the online reservation services—he favored one of the early ones that didn't apprise you of exactly where you'd be staying until you and the computer had settled on a price and a neighborhood. "It's amazing," he told me. "The rooms are so cheap."

"But what if you get there and they've never heard of you?" I said.

"They've heard of you," he said.

"What if they haven't?" I said. "What if you fly in from another city and they've never heard of you and they don't have a room?"

"You'll be doing this soon," he said. "Trust me. You'll get used to it."

I trusted him. And most things, I could get used to.

But not this. Not what we were going through now.

HE HAD ALWAYS BEEN THE ONE TO GIVE me advice. There was one time, though, just before or just after we were turning fifty, when he sought mine.

He was having money troubles. He was in a hole not of his own making, and he needed to get out of it.

"Do you think I should ask Chuck?" he had said to me.

It was an involute and wrenching decision for him to wrestle with. Acknowledging that a friend is in a different financial universe than you are is awkward enough—add to that the fact that the friend is not of recent vintage, but someone with whom you have been close all your life.

And then add this: The old friend is your brother-in-law. You're married to twins.

"I just hate to ask him," Jack had said. "But I don't know what else to do."

The embarrassment of asking aside, he was afraid that the request would somehow, in small but important ways, change his relationship with Chuck. And that would be awful.

"We're pretty far along in life to think that any of our feelings about each other are going to change now," I had said.

"I know," he'd said. "But I hate being in this position."

"I think you should do it," I had said.

I never asked him if he had gone to Chuck; I figured he'd tell me if he wanted to. I was thinking about it now because he had come to me looking for help with finding out about the nursing care. And I realized that he was a man who so seldom asked for anyone's help—who took such pride in solving all his problems himself, overcoming all his obstacles on his own—that it stood out when he asked. It was like an alarm going off.

NOT THAT HE'D EVER HAD A REASON TO hesitate asking anyone for anything. I couldn't imagine someone saying no.

In school, when students all over the building were

constantly trying to get away with every shortcut imaginable, wheedling to give themselves every leg up—from finding out what questions were on a test by asking friends who'd taken it the period before, to copying each other's lab book homework for chemistry class—no teachers ever suspected Jack Roth of trying to pull something over on them, simply because he wouldn't and they knew he wouldn't.

Parents always respected him—a rare enough thing. In our house, my dad was ever ready to show disdain for most of my friends, just on principle. A guy named Tim Greiner with whom I spent a lot of time returned one college vacation with a conservative and neatly trimmed new beard, and from that day on my father contemptuously referred to him solely as "Beardo." Once, after coming home with my mom from a too-long cocktail party, my dad actually took a swing at Chuck. I believe it was because my father had long viewed with disfavor that Chuck (1) consistently failed to stand up when an adult walked into the room, and (2) had "big teeth." On this particular night he evidently felt like knocking them out. (He missed. He may have intended to miss. It was an unusual evening.)

Jack, though, was always in a separate category. And he did his best never to ask anyone for anything. When he did, you knew it was not an entreaty arrived at lightly.

MY MOTHER TALKED TO JACK, AND SHE talked to Janice, and she got in touch with agencies and she checked with friends who had needed nurses in their homes for significant stretches of time. She found the information that Jack had hoped she would find; she made suggestions and offered to be a go-between and said that she would be following up. In short, she did a perfect job, and she did it quickly.

In telling me about it, she said something that brought me up short. I should have known, but I suppose I hadn't been allowing myself to think about it.

She explained to me about the various home-health-care options she had looked into—that was the phrase, bland enough, "home health care"—and then she said:

"And of course, after that he'll need hospice."

"Hospice?" I said. "He's not talking about needing hospice." Jack was walking around, having friends come over, speaking about driving again once the radiation-and-chemo doctors gave him the OK—what he was looking for was some in-home help later on. Hospice was for people at the very end of their lives—hospice, I knew from my own father's last months, was to help people die with dignity.

"That's not what he was asking you about, Mom," I said. "He doesn't need hospice."

"He will soon enough," she said gently.

And this was a woman who was not often wrong.

Thirteen

I TOLD JACK WHAT TIME I'D BE ARRIVING in Columbus and he offered to pick me up at the airport. I said he didn't have to do that; I said I'd check into my hotel and get myself to his house.

That was all right with him, he said, but he wanted to drive us to dinner tonight.

That was one of the confusing things about all of this. He was driving again just as he had vowed to do, he was going out for meals. But he'd had a CAT scan to see how the radiation and chemotherapy were working. Not so well; they weren't stopping what was happening to him, not the way the doctors had hoped.

"So what do you think, the Top?" he said when I called from Chicago.

I said the Top would be fine. Better to talk about that, about which restaurant to go to . . . that's what he seemed to be saying. There would be time later to talk about other things.

THE AIRPORT EYES WERE MOSTLY DIStracted eyes.

I could see that as I made my way from the arrival gate to baggage claim. That's how it seems everywhere; whatever mystique there once was to air travel—whatever faint echoes of refinement had somehow managed to survive the last fifty years—are gone, and probably for good. Airports, no matter how newly and expensively constructed, seem now at their souls to be Greyhound stations, but with government-mandated security, any giddy sense of moment long ago replaced by a grim and rote functionality. For many, air travel has become mainly a bother, seldom an event; the distracted eyes in terminals across the country are the eyes of weary people who in their minds are already, perhaps wishfully, at the next place.

My eyes were no different. I was hurrying through Port Columbus, wanting to get to Jack's house. There was a time when he and I would come out here just for fun. Fun—at an airport.

People—anyone, not just travelers with tickets—used to be allowed to stand on the roof of the original Port Columbus and watch as, several dozen feet away, the planes taxied from the gates for takeoff, or landed from far-flung cities before pulling up to the building to disgorge their passengers. Security? There was none. You could meander through the airport without anyone asking you a single question; you could stand on that rooftop outdoor observation deck, with a clear visual shot at all the planes, and no one thought it was out of the ordinary.

We'd do it all the time. It was one of the ways to pass the time in Columbus—watch the TWA and Eastern Airlines flights come in and out. The mother of some guys we went to school with—her name was Jerri Mock—decided she would like to be the first woman to fly around the world solo, and she did it, with Port Columbus as her starting line and finish line. We came out to watch her plane land; "Spunky Aviatrix," as I recall, was the way the headline in the paper described her the next day. Barry Goldwater, when he was running for president against Lyndon Johnson, made a campaign stop in Columbus, and we went out to see that, too. It had nothing to do with politics—we had none. It was that Port Columbus felt like the link between us and the wider world, and the sight of Senator Goldwater's, or Jerri Mock's, plane approaching, first a speck in the distance and then, as it grew nearer, something real, something about to alight . . . it made us feel that someday we, too, could be out there, way beyond the city limits.

There had also been the Three-C Highway, in the days before the interstates, but the Three-C, while undeniably a way out of town, lacked a certain grandeur. Probably the name had something to do with it. The Three-C Highway linked Cleveland, Columbus and Cincinnati, running north and south through towns along the way, all intersections and stoplights, no exit ramps; for kids in search of the promise of boundless horizons, Cleveland and Cincinnati didn't exactly make the heart race.

Once, restless, I asked Jack if he wanted to walk to downtown with me.

"From Bexley?" he had said.

I had said yes; I wanted to walk to downtown Columbus, and I wanted to walk back. I figured it would take an hour, maybe two, each way.

"You're nuts," he'd said. People didn't walk to downtown; they drove, or they took the bus. It was too far, and you had to cross Alum Creek, and, although there were no really dangerous neighborhoods along the way, it was just something that no one did.

"I want to do it," I'd said, and he'd declined, and so I said I'd do it myself.

It was mainly part of a wish to do something different. It was part of a wish to see things in a way I didn't see them every other day of my life. I walked downtown and I walked right back and when I got to my house and approached the front door my mother was waiting, concern in her voice.

"Where have you been?" she asked, and before I could tell her she said:

"Your Uncle Al called and said he had seen the strangest thing. He said he was driving home from downtown on Broad Street, and about halfway from downtown he thought he saw you walking on the sidewalk."

"That was me," I said.

"Why were you in that neighborhood?" she asked.

"I was just walking through it," I said. "I walked downtown and back."

"You *what*?" she said.

I'd just wanted to see some things; I'd just wanted to do something new.

Inside Port Columbus, my eyes one pair of distracted and unfocused eyes among many, I got on the escalator to baggage claim. I knew the way in the dark.

AT JACK'S HOUSE HE WAS SMILING AS HE greeted me and I could see right away that he had lost more weight. I didn't mention it to him, because I knew it was the last thing he wanted or needed to hear from me. And it's not like it was breaking news; he was well aware what his weight was.

He led the way to the living room. He was so proud of his house. In Bexley, there's a rough division of prestige according to where a person lives; south of Main is where the least expensive homes are, the homes where families first move and sometimes, if prosperity doesn't come, where they stay; north of Main but south of Broad is the town's middle-of-the-pack area, the area where my parents and Jack's parents lived when we were boys; north of Broad is where you live if you've really made it. It's an inexact equation; there are exceptions to the rule. But "south of Main" means something, as do the other descriptions; there is a shorthand everyone understands.

"The place is looking good," I said to him, and it was, the hardwood floors and gleaming kitchen and back patio made his house feel like a place you'd want to live for many, many years; he'd built it with that in mind.

"I remember when we had our first house after we got married," he told me. "Someone in my family said to me, 'You watch—your next house is going to be north of Main.'" Whoever said that had been right; with some intermediate stops along the way, he had made it to central Bexley, the same part of town in which he'd grown up. It had been a struggle for his dad and it had been a struggle for Jack, but here he was.

"The best thing is you were able to build it yourself," I said.

Not with his hands. But he had saved up so that he could have a house constructed from the ground up, and thus had presented himself and his family with the gift of forever being able to avoid the "old Woodruff house" syndrome. The town was so small, and everyone knew each other so well, that if you bought an existing house the chances were good that for generations people would refer to it not as your house, but by the names of the families who had lived there before. You lived in the old Woodruff house, or the old Monett house, or the old Hoffhine house. It provided a sense of community, yes, but there was also something confining about it, something cloistering. You were never quite you.

That, combined with an unspoken sense of reluctant real-estate competition with people we'd grown up with, helped propel Jack out of town, or so I'd always thought. As boys, I don't think we ever compared the relative costliness of houses; we didn't judge each other, or our friends, by our neighborhoods. If a kid was pitcher on the baseball team, that carried more weight than what his parents' property taxes were. But once Jack was in Bexley on his own, he couldn't avoid it; he may seldom have spoken about it, but if you stay in town you know that how you're doing is going to be publicly judged at least in part by where you're living. He had lived south of Main; Chuck, among others, had lived north of Broad. It was no great shock to me that, when a job in Minnesota beckoned, he took it. It wasn't just that the salary was better. It was that everyone he had ever known would not always be evaluating his street address.

But he was back now, Minnesota and an intermediate stop in California behind him, he was north of Main, if not north of Broad. He and Janice were the first people who ever slept in the bedroom he'd built; he and Janice and Maren were the first people ever to eat in the dining room. As the house was being erected he had walked me through it; bare beams and exposed pipes and unfinished walls aside, the place had felt like a grand estate to him even before the carpenters and roofers and plumbing contractors were halfway through. "This is going to be a sitting area," he had said as we toured the place during construction; even then, I could tell, it was a place that felt like his. Like home.

"We've got a little leak in one of the upstairs bathrooms," he said to me now. "And there's this creaking sound on the staircase that I'm going to have to do something about." He wasn't griping; if you can point out a house's flaws and somehow do it with a sense of pride, that's how he sounded. The place was lovely, and it was his, and he wanted it to be faultless. "I'm making a list of things that have to be fixed," he said. This was the house where he had intended to grow old.

"JAN, ARE YOU ON THE PHONE?" HE called out.

We could hear her talking in another room.

"Do you know how long you're going to be on?" he called. She didn't hear him; she kept talking.

"I've got to make a call," he said to me. "She's on the main phone—I'll just use the cell."

He reached for his cell phone, started to punch in a

number, then looked at me and said: "We could never have used these things if we wanted to call Dick Groat. How could we have both been on the line at the same time?"

It was the first thing we did together that made us feel cool, or close enough to it; it was the first thing to make us believe that on our own we had thought up something great. Looking back, I realize it was the first time we had felt that remarkable human stirring: ambition.

"I couldn't wait every month until we tried to pull it off again," I said.

We called sports stars. Real sports stars. He would pick up the receiver of one phone, I'd pick up the receiver of another—we'd switch houses in which to do it every month—and we'd make the calls. The one of us doing the dialing would have to yell to the other in another part of the house once the dialing was complete—you couldn't dial a phone if another extension was off the hook.

So one of us would dial, and then shout for the other to pick up—and when someone answered at the number we'd called, one of us (we worked this out in advance; we rehearsed it) would say: "Could I please speak to Jack Nicklaus?" Or "Could I please speak to Jerry Lucas?"

We were in seventh grade when we started this. Lucas was our first. He was in his initial year as an Ohio State basketball player—on the team that would go on to win the NCAA championship—and we decided (dreamed) that we wanted to interview him for the junior high school paper, the *Beacon*. But how do a couple of twelve-year-olds do that? We couldn't get into Ohio State basketball practice; we couldn't get into a postgame locker room.

We had read in *Sports Illustrated* he was a member of the Beta Theta Pi fraternity. One night at dinnertime we each drew a deep breath, ran to different rooms of Jack's house, looked up the number of the Beta house on the Ohio State campus, dialed, and when someone answered Jack said: "Could I please speak to Jerry Lucas?"

Amazingly, the fraternity member who answered didn't ask who was calling. He just yelled downstairs to the dining room in the chapter house: "Hey, Luke, it's for you."

Thirty seconds later, in a deep voice:

"Hello?"

We did it just like we'd practiced beforehand. We couldn't believe it was happening; we couldn't believe we'd gotten him on the line.

"This is Jack Roth," Jack said into his receiver.

"This is Bobby Greene," I said into mine.

"We're from the Bexley Junior High School *Beacon*," Jack said.

And we asked for an interview.

And—miracle—Jerry Lucas did not say no.

"I've just begun to eat," he said. "Could you call back in half an hour?"

Which we did, and we asked him five or six questions (taking turns—we never wanted to do this alone, the whole idea was that it was something we did together), and when we turned in our interview, at first the faculty adviser to the *Beacon* thought we had made it up. We had to convince her—and it wasn't easy—that we had really talked to Jerry Lucas, that this wasn't fiction. We might as well have told her that we had landed an exclusive interview with President Eisenhower.

After Lucas, we were hooked. We got Nicklaus, who lived in Columbus; we got Lucas's basketball teammate Larry Siegfried, we got Ohio State football fullback Bob

Ferguson. We would scan the downtown papers to see when athletic banquets were being held in Columbus (the Touchdown Club was the big one) and on the day of the dinners we would call the two major downtown hotels, the Deshler Hilton and the Neil House, knowing that the out-of-town stars being honored at the banquets would almost certainly be staying there. That's how we got Dick Groat, the shortstop of the Pittsburgh Pirates; that's how we got Bill Mazeroski, his second-baseman teammate; that's how we got the legendary retired Washington Redskins quarterback Slingin' Sammy Baugh. We never missed; once we'd set our sights, we never failed. "This is Jack Roth . . ." And they would talk to us. They would actually talk to us.

"You remember the big final question of each interview, don't you?" Jack said now.

"Of course," I said. We'd trade off each month; one month he would get to ask it, one month I would. It was always the same question:

"What is your advice for young athletes?"

"Can you remember what Larry Siegfried's answer was?" Jack asked me now.

"How could I forget?" I said. "*'Keep them grades up.'*"

"'Keep them grades up,'" Jack said.

Our stories would be typed and cut into mimeograph sheets, would roll off the mimeo machine smelling of that pungent ink, would take up a column or two in the next issue of the *Beacon,* and we were in heaven; twelve years old, and our classmates were reading our interviews with authentic sports idols.

"Quick," I said to Jack now. "How did we end every story?"

" 'We would like to thank Mr. Mazeroski'—or whoever it was that month—'for his wonderful cooperation.' "

"I'm not talking about the wonderful cooperation line," I said. "I'm talking about the last sentence in every story."

"Oh, you mean the sign-off," he said. " 'Yours in sports, Jack Roth and Bobby Greene.' "

We'd change the order of our names every month; we kept track. The whole thing was beyond intoxicating: see a sports star on TV, decide we were going to get him to talk to us, figure out where to call, and then do it, together. And each month, those tingling first few seconds when Jack would go to one telephone, I would go to another, we'd hear each other breathing as the phone rang on the other end . . .

And the roll of the dice, the leap into the unknown, the words we could barely conceive we were saying: "Could I please speak to Woody Hayes?"

"I couldn't sleep the night before every call," Jack said.

"I couldn't sleep for two days after," I said.

"Couldn't have done it with cell phones," he said. "We couldn't have been in on the same call."

"It wouldn't work now anyway," I said. "Can you imagine anyone putting two twelve-year-old kids through?"

"Do you remember what Jack Nicklaus said to us at the end of our call to him?" Jack said.

"What?" I said.

" 'Thank you for calling,' " Jack said. "Nicklaus actually thanked us."

"For our wonderful cooperation?" I said.

"Yours in sports," Jack said.

I FELL IN LOVE WITH IT; THOSE AFTERnoons with Jack, calling the sports stars, made me know that asking questions and writing stories was what I wanted to do with my life. That's how it began. I would read the Columbus newspapers and the national magazines that my parents subscribed to, and it's all I wanted to do. My favorite magazine was *Life;* that's the one I would read cover-to-cover the day it arrived every week.

Jack wasn't as smitten with the idea of doing it for a living as I was. He liked it, but it wasn't destined to be a lifelong romance. He never forgot, though, how strongly it had affected me; he never forgot that he'd been there with me, on the other phone line.

When we were in our early fifties my phone at work rang, and it was Jack, calling from a New York street on his cell phone. He said he was on the corner of Lexington and Forty-third.

There was a man on that corner, he said—a man selling very old *Life* magazines from a makeshift stand he had set up. The vintage *Life*s were going for twenty dollars apiece.

Jack was in New York on business. He had appointments to keep. And—knowing how much I had always loved *Life*—he said, over the miles: "What years do you want? I already had the guy give me the week you were born, but what other years do you want?"

"How much money do you have with you?" I asked.

"I can go to the bank," he said. "What years do you want?"

He was busy. But he stood on the New York corner and described the covers to me: "There's Donna Reed, from nineteen fifty-three . . . ," he said.

We spent ten minutes on the phone as he went through the old *Lifes*. And then, four hours later, he called me back. "All right, I went and got some money," he said. "I'm back at the guy's table." He was devoting a good part of his day to this. "There's Frank Sinatra Sr. and Frank Sinatra Jr. from nineteen sixty-three. . . . Here's the astronauts. Do you want that one?"

"Listen," I had said to him, "you're spending a lot of time on this."

"That's all right," he'd said. "How many times am I going to run into a guy selling the magazines you've been looking for?"

And in Manhattan he started up again, flipping through covers.

THE CIRCLE GOES 'ROUND. IN 1999 I BEcame the columnist for *Life* magazine—the monthly *Life*, the one that succeeded the weekly *Life* I'd grown up on. It was one of those things that, for public consumption, for others to observe, you try to appear to take in sedate stride, while inside yourself where no one can see it you're grinning ear to ear like the excited kid you really are and leaping into the air clicking your heels.

For the year-end edition that year, I was asked to do a theme-setting piece, summing up the century—talking about the transitional moment as the twentieth century ended, and the twenty-first began.

We went back and forth about what tone it should

take—how do you boil down a century into a magazine column?—and I persuaded the editor, Isolde Motley, that instead of trying to fill a big canvas, I should keep it as small as I could. To distill the century, I told her, I'd like to go back to the town where I grew up, to think there about a hundred years of changes.

So I went to stay with Jack. He had just moved into his new house.

The meaning of the trip—of sharing this *Life* end-of-the-century moment with him—was not lost on either of us. We had started out on two telephone extensions dialing our sports heroes, and because we had done that together I had found my life's calling; in that town of ours I had fallen in love with the great American magazine with the red-and-white cover logo, and here we were, at century's end, with me sleeping in his new house, on assignment for *Life*. It was a wonderful visit; Jack and I, together, talked long into the night about what the story should be and how to get it right, and the circle had come all the way around.

Life, in its monthly incarnation, did not last very much longer. When it died those associated with it moved on to other projects, other jobs. Everything has its own lifespan. Now, at Jack's house, we got ready to go out to dinner, and I looked on a shelf, and there it was, where he had saved it, the last *Life* magazine of the twentieth century, the one for which he had helped me think the story through. Things begin; things end.

"CHANGE FOR A PENNY?"

He grinned as we drove toward the Top. I hadn't heard that phrase in years.

"What made you think of that?" I asked.

"I think it was around here," he said.

The motel, on East Main Street outside of Bexley, was called the Forty Winks. The rumor . . .

"Do you think it could possibly have been true?" I asked him.

"You know what they said," he answered.

What they said was this:

If you walked to the front desk of the Forty Winks, and said "Change for a penny," the clerk would provide you with a prostitute.

"How could that have been true?" I asked.

"All I know is what I heard," he said.

Women were such a mystery to us—the idea of being in bed with a woman was so otherworldly—that of course, back then, the "Change for a penny?" story would seem rational. That a woman might in actuality sleep with you was such an incomprehensible, almost alchemical, concept that "Change for a penny?" made as much sense, in theory, as anything else.

Which is to say: We had no idea.

" 'Change for a penny' wasn't any less logical than that phone booth out on Cassady," I said.

"At least we knew that didn't work, because I tried it," he said. " 'Change for a penny,' we never tried."

The phone booth—on Cassady near the corner of Fifth Avenue, north of Bexley, past the railroad tracks—was where, it was reputed, you were supposed to stand if you wanted a woman to come by and have sex with you.

It was "Change for a penny?" without the necessity of any desk clerk interaction. You stood in the phone booth—so the story went—closed the door, did not pick up the phone, and eventually a woman would come by and tell you that she would sleep with you. It sounded like magic, which is probably why it appealed to us.

We tried it. The rest of us parked in Chuck's car while Jack stood in the phone booth at night. This wasn't the swellest neighborhood; being there at all at night was probably not such a good idea. But Jack had stood there . . . and stood there . . . and stood there. He, and we, had believed it. Stand in a phone booth, and a woman is yours.

"What would you have done if a woman had showed up?" I said.

"I don't know," he said. "She probably would have told me what to do."

No woman had showed up; we had gone home. Or to the Toddle House.

"It's sort of amazing to think we really believed a woman would appear just because you stood in a phone booth," I said. "Think of that now."

"Forget the woman part," Jack said. "What would be amazing now is a phone booth. When's the last time you saw one of those?"

THE EVOLUTION IN MEN'S LIVES AS THEY learn about women is one thing. Jack, in one lifetime, could progress from a kid with Forty Winks dreams, a lust-filled kid who stood in a phone booth, to a responsible husband and father . . . that transformation was tenable.

But . . .

"Salmon?" I said. "Scallops?" At the table, he had just said both words.

"Those would taste good tonight," he said. "I'm looking to see if they have them on the menu."

"How long have you been eating salmon and scallops?" I said.

"I don't eat them together," he said. "I eat one or the other."

"That's not the point," I said. "When did that happen?"

"I don't know," he said. "Years ago. Your tastes change. Everyone likes salmon."

"But there has to be a first time," I said. "How did it happen the first time—all of a sudden after a lifetime of regular food you decided out of nowhere that you wanted scallops?"

"You're driving me crazy," he said. "Let me read the menu."

This felt right; he'd told me, at his house, how tired he was becoming of everyone asking how he was getting along. He knew their motivation was benevolent; he knew they were only trying to express their concern and support. But all of this, he said, was having the effect of making him feel he was defined by his illness.

"That's not who I am," he said. "I'm the same person I've always been. But now whenever people call me or see me, it's like the one and only thing they want to talk about is my health. I hate it."

And on top of that, he said, he found himself having to repeat his various medical evaluations—what this doctor said, what that doctor recommended, what prescriptions he was taking—over and over. For each person who called asking to know how he was doing, he would go through the whole litany again—from initial diagnosis right up to the latest CAT scan results that had showed things were moving in the wrong direction.

"So tell them you don't want to talk about it," I said.

"I can't do that," he said. "I can't be rude to them."

"Then let me tell them for you," I said.

"Right," he said. "That's just what I need. You and your great social skills being my spokesman. I wouldn't have any friends left."

At the Top, we did our best to leave all that behind for a few hours. He was right: What had happened to his health wasn't who he was. It was just something that had fallen out of the sky. It wasn't welcome at this table. Not tonight.

"What's funny is how much people like spinach-and-artichoke dip," I said.

"Spinach-and-artichoke dip is delicious," he said.

"I know," I said. "It makes your mouth water just to think about it. And then you realize something."

"I give up," Jack said. "What?"

"Think about if someone had told you, when you were a kid: Here are the two things that someday are going to sound absolutely delicious to you, that are going to sound better than a candy bar. Spinach. Artichokes. You wouldn't have believed it."

"Can we just order?" Jack said.

"If they don't have scallops, maybe they've got some Brussels sprouts or broccoli for your dining pleasure," I said.

"I'm having a steak," he said. "You satisfied?"

"Let's get some onion rings on the side," I said.

WE HAD A GOOD EVENING, AND AS HE drove us back to his house just the fact of him being behind the wheel added to the delusory sensation that nothing in our lives had been altered, that this was all a storm cloud that would soon pass, that everything would someday be as it was. He was driving his car, wasn't he? He was laughing at the dinner table, and telling stories, and as zestfully involved as he had always been.

Yet I saw the grayness in his complexion, the bleakness beneath his eyes; he had pulled everything together for our night out, but he knew what the CAT scan had showed, he knew, and I knew, where this was heading.

At his house I asked if I could use his computer to check my e-mail. He said to go on into the room he used as an office.

He'd left the computer on, and the last site he had been looking at was still on the screen. It was a travel site, of the kind he'd first told me about years before—one of those sites that give you the best deals on hotels and flights. I smiled; another sign that maybe things hadn't changed so much.

I looked at the screen. He had selected a hotel in New York—a place called the Bentley.

Then I remembered. He was trying to get an appointment at the Memorial Sloan-Kettering Cancer Center. He was hoping to find a doctor who would tell him that there was hope—that the CAT scans weren't really saying what they appeared to be saying.

That's why the hotel was on his screen. That's what he had been looking at before we'd left for dinner.

"Greene, you in there?" he called. "You need help figuring out how to use the computer?"

"I'm fine," I called out to him, no longer wanting to spend any time looking for e-mail, walking back out to join him and plan for our day tomorrow.

Fourteen

YOU KNOW WHAT I LIKED BEST ABOUT the bookstore?" he said.

"I'm guessing it wasn't the money," I said.

"I liked the talking to people," he said. "People would come in and look at the books, and they'd ask me for suggestions, and we'd end up spending lots of time having really good conversations. Not just about reading—about all kinds of things."

He had founded My Back Pages in the 1970s, when doing such a thing—renting a storefront, stocking it with new and used books he thought people would appreciate, making it essentially a one-man operation—seemed like a fine business plan. He liked books, and even more than that he liked people who liked books—what better way to spend your days and nights than on something you truly enjoyed, in the company of people who enjoyed it just as much as you?

"Nothing made me feel better than when someone would ask me for a recommendation, and they would buy the book, and then come back after reading it and ask me to recommend something new," he said. "It was like they were telling me they knew they could trust me."

"You're lucky you didn't have Amazon to compete with," I said.

"I would never have opened the store," he said. "It was bad enough competing with the big chains."

The used books in the store, he bought from his customers and then put on the shelves and resold. But with new books—current bestsellers—he said he could always hear something in the voices of the book publishers and wholesalers when he placed his orders.

"I'd ask them to send me four copies of a book," he said. "And I knew they were sending four thousand copies to the regional distribution warehouses of the chain stores. Talk about feeling like a little fish."

One thing he didn't miss was all the Fridays when he'd add up the week's receipts and ask himself why he was even trying: "After a while you stop fooling yourself into thinking that one day customers are going to knock down the doors."

But sometimes he still did yearn for something else about the store.

"All those long talks with people who saw the sign and came in to look around," he said. "Sometimes someone would talk with me for an hour about books and then leave without buying anything. You'd think that would have made me mad. But it didn't. I would go home thinking about the conversation, and thinking it had been a good day."

WE WERE DISCUSSING THIS ON THE WAY to my parents' old house. He was wanting to walk every day; even though it wore him out, he continued to think it was good for his physical well-being, and he didn't like to do it alone. His doctors had told him it wasn't a smart idea to be exerting himself with no one alongside, should anything go wrong. Janice was adamant that he have company.

"You get used to a little privacy over the years, you know?" he said. We walked along Bryden Road. "Talk about not knowing how good you have it till it's gone—you never think about how much, as an adult, you appreciate the chance to be alone. And then no one wants you to be alone."

"You can be alone at home, can't you?" I said.

"I can be in a room by myself, but until this all stabilizes, I'm not supposed to be in the house alone," he said.

"It's hard to blame them for being cautious," I said. "Everyone remembers that first day when they found you on the floor."

"I don't even remember it happening," he said.

The house where I grew up was straight ahead, half a block away. "I'm serious about missing the option of being by myself," he said. "Think of all the times in your adult life when you really need your privacy. So many times, when I'd want to think about business, or make plans, or just set my mind at ease about something, I'd take a walk by myself or go for a drive or go have a cup of coffee somewhere and sit at a table by myself. I'd like to get that back. At least have the chance to."

In front of my house—there had been three owners since my parents moved out—we stood and looked up. "You know," Jack said, "it really wasn't a very big house, was it?"

"I loved this place," I said. "But, no, it wasn't real spacious. We were sort of right on top of each other."

"I think my parents' house was even smaller inside," he said. "But did you ever feel that you didn't have enough privacy in your house?"

I'd never thought about it. "I guess if I'd really wanted to be alone as a kid, I'd lock my bedroom door," I said. "But I didn't think my parents or my brother and sister were all over my business, if that's what you mean."

"Maybe the need for real privacy doesn't kick in until you're older," he said. "I've never liked a lot of noise around me in the office. I don't like to work if there's a whole lot of talking going on. But don't you figure our houses were full of voices and noise when we were kids?"

"We didn't have anything to compare it with," I said. "It didn't sound noisy because we'd never spent any time anywhere where there weren't voices all the time. We didn't know any better."

"That's what it's like for me now—voices always around," he said. "I'm not saying I'd want privacy all the time. But it would be nice to have the choice."

OUR GOAL, ON THIS DAY, BECAME MAKING it to the track.

He was fatigued; I could see it even as we stood in front of my family's old house. We hadn't been walking long, but he simply didn't have the stamina.

"You want to just head back home?" I said to him.

"No," he said. "No. I'm good for a little while more."

So the track it was. Every time he and I walked, he told me, no matter where else we might go, he wanted the track, and his old house, to be part of the route.

We walked up Roosevelt toward Fair, where we would turn to get to the track, and even this—the fact that we had a specific goal in mind, that making it to the track was our objective—was a signal of how much had changed so quickly. All the thousands upon thousands of days and nights we had spent in town, as children and as adults, and we seldom had a goal when we went out—we seldom had a destination. The whole idea was that there never was an end point. Not one that we were looking for.

Someone driving the opposite direction on Roosevelt honked his car horn, and Jack waved.

"Who was that?" I asked.

He said a name I didn't recognize. Someone who had come into his life in the years since I'd left town.

"That's the nicest thing about living here," he said. "How many places are there that people would honk to say hello every time you went out for a walk? There's never a time that doesn't happen to me."

At the track he said "Take it slow," as if I didn't know. When we were parallel with the north end zone of the football field he said, "All those Friday nights, when it seemed like what went on on this field was the center of the whole world."

"There was one Friday night, when Dave Frasch hit Tim Greiner for a touchdown in the corner of the end zone," I said to Jack. "Right . . . there." I pointed to the place where Greiner had caught the ball. "At that moment, I doubt there was anyone anywhere in the world who could have felt any greater than Dave and Tim felt. The people in the stands were going crazy."

"When the crowd in this place cheered, you could hear it all over town," Jack said. "Even when you were a little kid, too young to go to the games, on Friday nights in the fall there would be this sort of soft rolling echo, that made it all the way to your house, and you knew."

"And there was that haze from the stadium lights," I said. "It would float over the town, like a cloud."

I looked at the patch of the end zone where Greiner had reached up for the ball. We didn't know, then, that this wasn't the goal line at all; we didn't know that there were endless goal lines after this one, that once you left this field and left this town you were going to be confronted with one goal line after another, sometimes not of your own choosing, that as soon as you reached one it would disappear, to be replaced by another off in the distance. Often it seemed that you'd never get there.

The scoreboard at the end of the field was blank. Jack had a new scoreboard now—blood counts, and radiation levels, and carefully measured doses of the harshest varieties of medicines, things we'd never thought about, things I wished we'd never had to. Sometimes on Friday nights the home team that played here won, sometimes they lost. It seemed of high significance, at the time.

"YOU'RE NOT REALLY SUPPOSED TO USE this gate, but it's open, so I guess we can," he said.

"What, are you afraid Officer Butters is going to come by?" I said.

We left the track and went through the gate and over to Stanwood Road.

"Standing lookout and watching for Butters made me feel like I was in a James Bond movie," he said.

"Good guy or bad guy?" I said.

"That depends on whether you think Butters was the good guy or the bad guy," he said.

It was about as anti-authoritarian a thing as there was to do in our part of the world back then: in the dead of night climb to the roof of the elementary school and pull the hands off the big four-sided clock, or cut the chain and lock off a gate that was supposed to keep us from taking a shortcut to school, while one of us stood on each end of Stanwood so that we could yell a warning if Butters's police cruiser came into sight.

"I don't think he ever knew any of our names," I said.

"He wasn't supposed to," Jack said. "That would have made him seem too human to us. He had to maintain that distance."

He was a thin, taciturn young uniformed Bexley police officer who was assigned to the night shift, and who would cruise slowly through town, just checking things out. He never spoke. We knew virtually nothing about him except his name. Crime was not a factor in town; there had not been a murder, or so we were told, in all of the twentieth century.

"How many police officers do you think there are in Bexley?" I asked Jack.

"I know how many there are," he said. "Twenty-eight. Divided among three shifts. It was in the weekly *Bexley News*. How many police officers are there in Chicago?"

"About thirteen thousand," I said.

"There are as many cops in Chicago as there are people in Bexley," he said.

We turned on Elm toward his old house.

"You think Butters ever felt scared?" I asked.

"About what?" Jack said. "There were no criminals to catch."

"I don't know," I said. "Just about driving around alone, knowing that if anyone ever figured out that he was just one guy on his own, he'd be a sitting duck. A cop like Butters in a town like this, most of his power comes from people not really thinking about how totally outnumbered he is. If he'd ever been surrounded by people wanting to make trouble, I don't know how much backup he'd have had."

"It did seem like he was the only cop out there most nights, didn't it?" Jack said. "He didn't have a partner, he never stopped to make conversation with anyone . . . he just cruised up and down the streets."

He paused to catch his breath.

"What was Butters's first name?" he asked.

"We never knew," I said.

"If we had, it might have blown the whole mystique," he said.

THERE WAS A *THWACK* SOUND, THEN A silence. Then another *thwack*. And again.

"Over there," Jack said.

A kid was throwing a tennis ball against the side of his house. It bounced back onto his driveway; he caught it, paused, then threw it again.

"Has there ever been a time, when we walked through town, when we haven't seen that?" I said.

"Maybe not in the dead of winter," he said.

We used to do it ourselves. It was an old sight to us now—we really had, in our twenties, our thirties, our forties, our fifties, seen successive new generations of kids doing the exact thing. It was as if they had become us—but they didn't know it.

"Look at his face," Jack said. "It's not old to him."

And it wasn't. There was excitement—and determination—in the kid's face. He was pretending to be someone—a baseball player, a football player, there was no way to tell. The tennis ball had become some other kind of ball, in his mind; his driveway had become a big-time arena.

"Do you think there's any part of town where we've never been?" Jack asked.

"I doubt it," I said. "I can't think of one."

"That's what was the best thing about when we first knew each other and walked around," he said. "Every block was someplace we'd never seen before. We didn't know what was around the next corner. It almost felt like we had to have permission to go from one block to the next. Like we should have had a visa."

"The town's never going to get any bigger," I said. "It's surrounded on every side. If there's nothing we haven't seen yet, then there's never going to be."

"What do you think we did to replace that feeling?" he said. "Once we'd seen every single block and every single house and every single store in Bexley, what did we have to look at that we hadn't seen?"

"The world?" I said.

He laughed. "Other than that," he said.

The tennis ball *thwack*ed against the kid's house, it bounced back to him, and he caught it. In his mind he was in the packed arena somewhere, but he was safe at home. Not a bad combination. We knew. We'd been there.

WHEN HE WAS IN HIS THIRTIES AND FORTIES, when there were signs that Bexley might not be as safe as when we'd been growing up, Jack worried about it.

There would be police-blotter listings, at the time, in the *Bexley News,* and the ones that had bothered him the most were the ones about bike robberies. He would cut them out and send them to me.

"Is this awful?" he would write on the clippings. They usually were on the same theme: A child in Bexley would be riding his bike, would be approached by an older teenager who had come in from Columbus; the older boy would demand the bike and ride away on it. One I remember especially, because it had been vivid and Jack had been particularly disturbed by it. A boy had resisted giving up his bike; the robber had slugged him in the face before taking the bike and riding away. The boy had run home in tears.

"I don't know what to make of what's going on," Jack had said to me at the time. "This is something new."

He was sending the clippings to me in a town where there were more than five hundred murders most years; he was sending the clippings to me in the town that had thirteen thousand police officers. He knew that. Yet he also knew that I would not shrug off the meaning of the bike robberies back home.

"It's not so much that the bikes are being stolen," he told me once. "If they were stolen out of garages, or back yards, that would be one thing. What upsets me so much is that the robbers are going up to the kids on the street, while the kids are riding their bikes, and telling them that they have to turn them over.

"Think what that does to a kid. He stops trusting people he doesn't know. That's going to stay with him—that's going to change him. The world becomes a place where people take things from you just because they can. And the parents are telling the kids to just give up the bikes if they're approached. The parents don't want their children beaten up, or even worse, over a bike. But think about that lesson. If someone wants something that's yours, just hand it over, or you'll be hurt. You don't think that changes how a kid looks at the world?"

I couldn't disagree with him; maybe he was lucky to live in a town so peaceful that the theft of a bicycle would make the local weekly paper, but I knew he had not been wrong to be troubled.

The bike robberies seemed to have stopped, or at any rate diminished in number; either that, or they weren't being reported in the paper any longer.

Or—a worse possibility—they were being reported, and Jack had become so accustomed to them that they didn't upset him anymore, that he skimmed right over the police-blotter listings and no longer clipped them out to send to me.

That, I didn't believe. There are people who can become hardened, but for better or for worse he would never be one of them. The *thwack* of the tennis ball against the house grew fainter, we walked farther toward Ardmore Road, and within a minute or two we were again at Audie Murphy Hill, named so long ago, by us, for America's fiercest man of combat.

"I JUST WANT TO STAY AROUND FOR JANice and Maren," he said.

It's the first he had brought it up directly all day—the idea of what might be coming.

"I know," I said. "And they know."

"When I think about them being around, about them being in the house, waking up every morning, and I'm gone, I'm not in the picture . . ."

I looked at his old house, and I remembered certain nights after his mother had died, when his father would be downstairs, and Jack and I would be up in his room.

There was a sound—I could hear it in Jack's room, drifting up the staircase from the living room.

Mr. Roth had an early version of a remote control for the family's television set. It wasn't like the ones today—it couldn't shift instantaneously and inaudibly from channel to channel.

It was clunkier and more mechanical than that. This was in the years before cable television—there were only three channels in Columbus, four if you counted the educational channel. And what the remote-control device would do was noisily move the channel changer on the TV set around its axis.

Mr. Roth didn't have much of a social life. Most evenings he was just in the house. And from that darkened living room, lit only by the glow from the TV screen, I could hear it up in Jack's room. The loud, metallic *clunk, clunk, clunk*. Irvin Roth would sit there on the couch in the shadows, where he and Mrs. Roth had once talked and laughed and had family gatherings and lived their married life together . . .

He would sit there, pushing the button, and the noisy channel changer would move from Columbus's Channel 4, to Channel 6, to Channel 10, and then around to Channel 4 again. He would search for something to fill his evenings, before going to bed to rest before another day at the fruit-and-vegetable market. Jack and I would be hanging around upstairs and it would always be in the background, the sound of Mr. Roth seeking something. I don't think it was until years later that I thought about the loneliness of the sound, the sound of a surviving spouse attempting to get through one more night.

"How long have we been walking?" Jack said.

"About half an hour," I said.

"Boy, it seems longer," he said.

He looked at me.

"I just can't believe this is happening," he said. "We've always talked about everything. But we never talked about it ending."

Fifteen

One morning I arrived at his house and I could see that Janice was in good spirits, which could mean only one thing: Jack was in better spirits, too.

"Go on back and see him," she said. "He's done some business already today."

He was in the living room, his socked feet up on the coffee table, and he had a look of accomplishment to him.

"I sold five trucks of candy today," he said.

"And you know I have absolutely no idea what that means," I said.

"It means I can pay some bills," he said.

"But how do you sell candy by the truckload?" I said.

He explained. He'd had the private-label candy available to him in a warehouse somewhere, he'd found a buyer at one of the national discount-store operations, he'd acted as the go-between, and the five trucks—big semitrailer truckloads, as I understood it—were on their way. He was paid a certain amount for each truck of candy.

"You did it from right here?" I said.

"I was working on the computer since early this morning," he said. "Between the computer, and my office phone forwarding to the house, it's like the computer and the phone are two employees."

"So if you can sell five trucks of candy every day, you're in great shape," I said.

He grinned and shook his head, as if I couldn't be more dense. "You don't sell five trucks of candy every day," he said. "You don't sell five trucks of anything every day."

WHEN HE WAS IN HIS FORTIES AND MADE the move from Minnesota to California to work for some sort of grocery wholesaler, I could tell almost immediately that something was wrong.

I didn't know exactly what; it was just the sound of his voice. It wasn't so much his words over the phone as the tone just beneath those words.

And it occurred to me that, even if he had elected to tell me what was amiss at his workplace, I would not have understood the details. I realized I didn't truly know what he did, specifically, for a living; I had always known the names of the companies for which he worked, and, in broad strokes, what line of business those companies were in. But precisely what he did—how he spent his days, the subset of skills for which he drew his salary—I was pretty much in the dark about that.

It was this impossible thing—at least it would have seemed impossible when we first knew each other: that something so central to his life would be something of which I was in many ways unaware. But it happens; all of us go on to choose which compartments of the work world we'd like to inhabit, and once we get there we don't think to tell our oldest friends what, brick by brick, our compartment is like. Maybe we fear they wouldn't be interested.

Yet if what he did with each quarter hour of his business day was something I didn't know, there was something I did know, and had always known, with complete certitude: the sound of his voice, and when something in that sound was awry. From his voice I knew he wasn't happy in California, knew something was not right, and I was not surprised at all when, soon after, he told me he and Janice and Maren were moving back to Columbus. But I couldn't shake the realization I'd had—his work, in many ways, was the core fact of his adult life, and I didn't know, didn't genuinely know, how he did that work.

IN OUR EARLY TWENTIES, WHEN WE BOTH were working in the Chicago area, we didn't see each other so often. I was starting as a reporter at the *Sun-Times* building on North Wabash downtown, he was teaching school in the suburbs, and there just wasn't much time.

On those weekends when he would come to the city to see me, or I would travel north to see him, we tried to act as if we were still on the same path together, as ever, but we knew we were fooling ourselves. It was one thing when we had attended different colleges in different cities—we might not have spoken as frequently (a long-distance call in that era was not something to be undertaken on a whim; friends at separate schools actually wrote letters to each other and dropped them in the mailbox), but still, we each understood that we were going through substantially the same drill every day, just on different campuses in different classrooms with different professors and friends. Our contexts were parallel.

Once out and drawing paychecks, though, we were a little adrift. We'd see each other and for the first time in our lives there would be pauses in the conversation: gaps. He would spend his days lecturing students in a suburban classroom, I would spend mine running around Chicago in pursuit of stories, and there was something in the air when we got together, something foreign and unwelcome. It was measured in those short silences. We were veering off on divergent tangents. We'd never felt that, in the company of each other.

And it was still too early in our adult lives for us to confess to each other that we didn't like it. For all we knew, a person was supposed to love this part of his life—the bidding farewell to all that had gone before. I was living in a small studio apartment in what was known as a four-plus-one building in the Rogers Park neighborhood—four floors of residences, the first floor a garage—and I had rented barstools for furniture, I knew practically no one in Chicago except my new newspaper colleagues. I was getting my byline on the front page, but who saw it? No one I cared about.

There were winter nights when I fantasized about the possibility of keeping the *Sun-Times* job but commuting from Ohio. If I went straight from the newsroom to

O'Hare after work each evening, caught the TWA non-stop to Port Columbus, slept in Bexley and then got the 7:00 A.M. TWA back to O'Hare . . .

It would cost me more than my entire salary, I calculated, but in the fantasy I could have back that which I treasured, and still pursue the job I knew was right for me. In my nonfantasy world, the real one in which I was living and working, I would see Jack on those Chicago weekends and we would talk about where Allen was, and Chuck, and Dan, and what I wanted was for all of that not to have ended. I sensed that Jack shared the same longing, but at twenty-two to admit that would have sounded like a confession of failure, so it went unsaid. And I felt that emptiness even as I was doing work that I thought I wanted to do for the rest of my life; Jack's work was not something he intended to stick with, so he was doubly unanchored, and in his face I would sometimes see the question: Where am I going? He would look off somewhere for an instant, and I knew what he was wondering, and I didn't have the words to tell him I understood.

THE IDEA OF GOING OUT INTO THE BUSINESS world—whatever our businesses might turn out to be—was not one that came especially easily to us anyway. With Jack, I long suspected that it had something to do with the first business he ever saw up close.

When he and I, as boys, would go down to his dad's market on Town Street, it would have the feel of a lithographs-and-high-button-shoes Old World, while being situated right in the business district of a twentieth-century American city. The market is gone now—replaced years ago by traditional center-city office structures. But in the early part of the 1950s, the downtown produce markets where Mr. Roth and those in his same line of work plied their trade might as well have been a part of the 1850s. Shadowy inside, brick-and-timber-framed, with canvas sacks of potatoes and vegetables strewn about—you expected to turn your head and see horse-drawn carts clopping up the road. Yet just to the west and to the north, a few short blocks away, was the AIU Building, Columbus's skyscraper. Jack and I would wander around his father's market, then step outside to look at the Columbus skyline, and it was like living in two worlds at once, like being a part of the dust-coated past and a part of the Technicolor-and-VistaVision present, which I presume was a feeling not unfamiliar to Mr. Roth himself.

By the 1960s Mr. Roth and the other produce vendors would move east to a sterile and amorphous series of newly fabricated concrete warehouses and loading docks near the airport; the operations there would be more efficient, the farmers would have quicker access and would not have to negotiate downtown traffic, any aura of past centuries would be erased. Airport-area motorists would not even have cause to glance over at the new outlying market buildings as they zipped past; the new markets blended into their featureless environment like cement camouflage, the new buildings were like immigrants who had at last lost their accents, and pretended to be proud of it.

But Mr. Roth's original market—the market on the edge of downtown—was where Jack and I first beheld the world of commerce, where we first observed what it looked like when a man went off in the morning to make a living for his family. We felt like small-town boys in that market; it was as low-kilowatt as a cave in there, it smelled like burlap and dried oats, and maybe, somewhere in our heads, we always thought that this, or something like it, was what awaited us when we became the ones to go to work to provide for our families.

"THE TRUCKS SHOULD BE AT ALL THE stores with all the candy within seventy-two hours," he told me. "Probably within forty-eight."

I had asked him, in his living room, about those five trucks of candy. How fast, once he had secured the orders, did the merchandise arrive at its destination? Once he had promised delivery, how soon would that promise be fulfilled? Very soon, he was telling me; the candy would get to where it was going within days.

He never had wanted to be a fruit-and-vegetable man. He knew that early. That was his father's world, and once Jack was out of college he learned that he did not aspire to be a part of it. He'd tried it for a while, after leaving the teaching job. His brother, Benson, had taken the produce business over from his dad, and Jack had gone to work with Benson, but it didn't take, his heart wasn't in it, and he had walked away.

"They tell me that I don't have the killer instinct," he said to me once in those days when he was first trying to find a place for himself in business. "I talk to people about coming to work for them, and they say that in order to succeed you need the killer instinct."

The businesses into which he was inquiring were ones in which rapid-fire deals were made, where a penny or two per liquidated-merchandise lot, the penny or two shaved under pressure from an offer, could mean the demarcation between a profit and a wash. It was the kind of business at which Chuck's dad had excelled and prospered; Chuck had followed his father into that kind of work.

"I don't know if I can learn to have the killer instinct," Jack had said to me in our twenties. As if lacking it was some sort of failing; as if not having the killer instinct made Jack himself damaged goods.

"What you have to ask yourself is why you'd want to," I'd said to him. Not having a killer instinct was the best thing about him; lacking the killer instinct was what set him apart.

But he had ended up, in a roundabout way, in some cranny of the same business his dad had been in after all; he may not have been down at the market before sunrise, stepping between cardboard boxes of onions and carrots, but he was moving edible merchandise, he was sending food from producer to consumer. Whatever the killer instinct was, I knew he still didn't have it—I'd met enough people who did. "Five trucks isn't bad for a morning's work," he told me, and maybe the buyers had bluffed or bullied a better deal from him than they would have elsewhere, maybe they had beaten Jack out of a penny per carton here and there. Maybe that was it. If so, he didn't seem defeated. He was having a good day.

"JACK, YOU BACK THERE?"

It was a man's voice. Jack stood up; he recognized it, although I did not.

A fellow I did not know—about our age—came into the room from the front of the house.

"Hey," Jack said. "I lost track of time."

"I may be a little early," the man said.

He was a friend from Columbus. Jack introduced us, we shook hands, and the man said: "The others are on their way. Show me where it is."

Jack led the man down to the basement. There was some work that needed doing down there—something about shelves, or cabinets—and Jack did not have the physical strength. So his friends had volunteered to do it for him.

I'd seen it over the years, on trips back home—Jack, as was to be expected, had good friends who had come into his life relatively late, and who meant a great deal to him. It was always a little jarring—he would be in a group of people I barely knew, they'd toss off catchphrases that meant nothing to me yet volumes to them, they'd laugh at references to memories I was not a part of and did not understand. It happens to everyone; if Jack had come to Chicago and found himself in a gathering of people I'd gotten to know through my work in the last twenty or thirty years, he'd go through the same thing.

Still, to see for yourself your oldest friend's new friends is always an instructive moment. And not only friends. In business, there are people who become important to you—people you would never select to be your friends, people you would never want to be important to you—and there's not much you can do about it.

Jack, in his thirties and forties, would tell me about bosses or partners with whom he was having some friction, and I would know that there's no escaping it: We all at one point or another find ourselves in the position where someone else has the power to evaluate us and tell us what we're worth. In business, it doesn't matter whether we agree with their evaluation—often we vehemently disagree. At some echelons of business, that's just about the sole job of certain executives: to tell other employees what they're worth. Jack, I knew, had to deal with the ramifications of that more than once, and at the hands of people about whose own worth he had deep doubts.

That can make for some dismal days in a man's life. But then there are days like this one—like this late morning at Jack's house.

"Janice, are they down in the basement already?" I heard a new voice say.

Another man; another face I did not know.

"They're waiting for you down there," she said. "I think they've already started."

They were coming to Jack's house; these friends of his, friends who had found him in the years since he and I were boys, were coming to do the work that he no longer could do. This one descended the stairs to the basement and I heard Jack greet him with a warm and energized hello.

Bosses come; bosses go. In the end, as often as not, they have to face their own sets of evaluations, rendered by people who may consider them every bit as disposable as all those upon whom they have passed stern and unmagnanimous judgment.

I could hear it from the basement: "Jack, go on back upstairs. We'll get this done. We'll have you take a look when we've finished."

These men had decided upon Jack's worth years ago; I didn't know them, but I knew that. They had evaluated him, and who he was, and here they were, without being asked. I could hear them hammering something down below. I wondered if they would have been quite so quick to come help a man who possessed the proper and perpetually prized killer instinct.

ON A TRIP TO COLUMBUS WHEN I WAS IN my forties, I was staying at the Hyatt downtown—a soaring high-rise from whose windows I could look down upon the block where Irvin Roth's market used to stand—and Jack, after his work was finished for the day, met me in the lobby so we could go to dinner.

"You know who works in the front office here?" he asked me.

I waited for the answer, and for a second the name he told me didn't ring a bell.

"Come on," he said. "You know. The lead singer for the Dantes."

The Dantes had been the top local rock band in Columbus in the middle of the 1960s; guys about our own age, they would compete in periodic Battles of the Bands at the Northland Mall, a new-at-the-time shopping center out on the far stretches of the interstate.

We'd drive up to Northland to stand in the parking lot and watch all the local bands go up against each other. One in particular, called the Marauders, was pretty good, but the Dantes had a lead singer who was the closest thing central Ohio had to Mick Jagger. Lithe, skinny, pouting, elusiveness in his eyes, he had that quality all of the best lead singers have: He seemed to come from somewhere far away, or even nowhere, he seemed to reside inside the songs instead of the world the rest of us occupied. We knew which high school he went to; we knew that, on paper, he was no different from any of us. He must have listened to the same radio stations we did, he must have watched Jimmy Crum broadcast sports on Channel 4, he must have shopped at Lazarus. But—that lead-singer difference, that can't-be-taught characteristic that defines the best of them—onstage he had the power of personality to make his audiences forget all of that. Onstage (or, in the case of the Northland parking lot, on pavement), he was a man from out there in the rock-and-roll ether, as separate as a spaceman.

The Dantes made a 45-rpm record, "I Can't Get Enough of Your Love," that all of us predicted would go Top Ten nationally, but it never quite broke out of Columbus. I hadn't thought about Northland, or the Battle of the Bands—never mind the Dantes—in many years. Until that night at the Hyatt when Jack had told me the singer was working in the executive offices.

"He was mentioned in *Columbus Monthly,*" Jack had said. "He's in the hotel's sales department or something. They had a picture. Coat and tie."

The Dantes, we had thought, were on their way—we didn't think it, we knew it, standing in the Northland parking lot, watching the singer skitter across the asphalt with his microphone in hand, we knew they were so talented and so good—so sharp, according to Jack—that they were pointed straight for the heights. And somewhere in our heads we must have believed that we were going with them. On a hot summer afternoon near the freeway, with the amplified notes from three guitarists ringing in your ears, you can buy that possibility: This is your town's best band, and they're heading places, and they're going to take all of us along for the ride.

Northland Mall is no more—torn down—and on that day in the Hyatt lobby Jack had told me about the lead singer in the front office. He had a family and the job was one he liked and he was quite pleased with the way things had turned out—that, Jack told me, was what the *Columbus Monthly* article had said about the man who had been lead singer of the Dantes. From my hotel room I could see where Mr. Roth had once sold his fresh fruit and vegetables, at least the piece of land where his place used to be, and things work out the way they're supposed to, or so we are always told.

CHUCK CAME OVER TO JACK'S HOUSE while the friends were working on the shelving in the basement. Jack went down to see how they were doing, and when Chuck and I were alone I told him some of the things I'd been thinking: about how I knew that some of Jack's previous companies had taken advantage of him, had treated him in ways he didn't deserve to be treated, had surveyed his trusting nature and had used it against him. Jack would never put it in quite those terms, but from what he had said to me over the years, I knew it was true.

"I think he's been happiest since he started his own business," Chuck said. "If there's a deal he's working on, and it looks like someone is going to get hurt, or be treated unfairly, he walks away from the deal. It's not worth it to him to have bad feelings with people he's negotiating with. You wouldn't think, with that attitude, he could make it in his own business. But he has. He's done pretty well."

I told Chuck about the killer-instinct conversations—the

ones about which Jack had informed me years before, when people were telling him he didn't have the requisite voraciousness to ascend to the wholesale-trading summit.

"I was a part of some of those conversations," Chuck said. "I don't think anyone was trying to insult Jack. It was more like they were trying to warn him. He'd been a schoolteacher and he'd run that bookstore. I think they were trying to tell him what it was he was getting into."

"You don't think he picked up any of the killer instinct?" I asked.

"Please," Chuck said. "Think who you're talking about."

ONCE JACK WAS IN CHICAGO FOR A TRADE show of some sort—a convention out at McCormick Place, the enormous exhibition hall next to Lake Michigan south of downtown.

He was one of tens of thousands of attendees—all of them in Chicago in the hopes of buying or selling something, or of meeting people who could help them do it later.

He had called and asked if I had time to meet for a drink when he was finished at the trade show for the day; he had dinner plans later with some other conventioneers, but he had time for a quick beer or two.

So we met in a downtown bar and he walked in—that smile, that chipped tooth, that nod—and I noticed something right away:

He had a name tag on. He'd forgotten to remove it when he left McCormick Place.

And . . . I don't know, there he was, my oldest friend, and

the name tag was color-coded and it said "Jack Roth," and the name of whatever company employed him at that point in his life, and underneath that either "Vendor" or "Buyer" or whatever his assigned role was that week, and . . .

It shouldn't have bothered me. It was stupid that it bothered me. But all I could think was, That's not who you are—no one should define you like that. Everyone at the convention wore those name tags, I'd worn name tags many times myself, and there was nothing so bad about what his name tag said.

But I thought, that night, of who we had been—those two kids riding the Bexley bus, tossing the socks at the basket above his bedroom door—and we didn't know, back then, what was waiting for us, we didn't know who we were going to become, and it happens to everyone, it happens to the lead singer of the Dantes, everything changes, you can't stop it. It's what the world does to all of us, as it funnels us into life's ever-narrowing chutes.

I knew it was dumb—I knew I was wrong for feeling bad about it. It was just a name tag, after all. But there was Jack with his name tag, and I wanted him to throw it away, I wanted him to toss it and for us to hop onto a bus out of Chicago en route to nowhere and to everywhere. When we were first going to school, when we were first friends, we would write those made-up song lyrics for our classmates to sing—song lyrics about geography, about spelling—and we would take the biggest hit songs of the day and reject the words that were in those songs, keeping the melodies but replacing the words with lyrics of our own. The words to that Tennessee Ernie Ford song: *You load sixteen tons, and what do you get? Another day older and deeper in debt. . . .* We changed those. We knew we could do better.

THE FRIENDS WHO'D BEEN HELPING JACK came up from the basement to take a break. I could hear him and them talking in the kitchen.

"Five trucks . . . ," I heard Jack's voice say.

The candy was on its way to the stores. Five truckloads sounded like a lot, but it's a big country, full of constant hunger.

Sixteen

I don't know if it was our talk about business that did it—I have a feeling that it was—but one afternoon, out of nowhere, Jack said: "Do you want to go over to Main Street with me?"

I told him that I'd be glad to go for him—Main Street was Bexley's business district, it had hit some less-than-booming times in recent years, but it was still where people went when they needed to buy something and didn't want to drive to downtown Columbus or to the distant malls. So I assumed Jack wanted to make a purchase on Main Street, and I said I'd save him the trip.

"There's nothing I really need," he said. "I just feel like going to Main Street for a while. You want to?"

The news from his doctors had not been so good, and every time he brought up something like this—something that once wouldn't have caused me to have a second thought—I asked myself if this might be the last time, and if that was the point of the request. I kept getting the feeling that Jack had an unwritten checklist: things he wanted to be sure to do one more time.

"We can go to Main Street," I said.

"Good," he said. "We should."

THE BEST I CAN DESCRIBE IT—THE BEST I can try to explain what was going on—is that he was trying his hardest to taste some things.

Food itself no longer tasted right to him—like mush, he had said that night at Chuck's for dinner; the radiation and the chemotherapy had ruined his sense of taste and had made anything he ate taste like mush.

But the tasting he was in pursuit of now—on our afternoon over on Main Street, as with just about everything else he and I were doing and would do—was of a different type. He was tasting his life. He was savoring who he was, and where he had been, who he had known . . . he was tasting it with a fierce and pervading kind of appetite.

It wasn't nostalgia; this was much more profound than that, this, in my eyes, was something that bordered on holy. All these months, instead of making them about death, he was making them about his life. And I was finding that it was the honor of my own life just to be alongside him.

"I can't even believe Norwoods ever existed," he said on Main Street.

"Now, we're not walking down there," I said. "That's way too far on foot."

He looked to the west. "I know," he said. "But the idea

that Norwoods actually was there, five minutes in a car from us . . ."

Norwoods was an amusement park, and—wonder of wonders—it was a neighborhood amusement park, right outside of Bexley on Main Street, on the way to downtown Columbus. A Ferris wheel and a steam-engine train ride, a merry-go-round and a shooting gallery, a haunted house, and little boats chained to guide rails beneath the water's surface in a murky man-made pond, a penny arcade with a row of wood-surfaced Skee-Ball machines . . . this was before Disneyland, before Great America, this was when someone must have had the idea that for the men just home from World War II—the men and their wives starting to raise families—an amusement park would be just what was needed.

And our dads would indeed take us there. It was the shortest of car rides—a few minutes down the street. Calliope music and the sounds of dozens of metal gears on dozens of rides turning at the same time, children lining up to steer miniature bumper cars inside an oval-shaped open-air enclosure . . . it was purely a local operation, it did no advertising that I remember, it was just a part of life in our town. A home-owned neighborhood attraction.

"I don't think it was even there for very long," I said. "I bet television put it out of business."

"You look now at where it was, and the lot is so small," Jack said. "All the rides I remember, I don't see how they fit them all onto such a compact lot."

There's a freeway entrance now, right where the edge of Norwoods used to be; unless you were alive when Norwoods was alive, you'd never guess it was ever there.

"Our dads would drive downtown to work, and they would pass an amusement park every day," Jack said, shaking his head.

"Not a bad sight as they started their days," I said to him. "If you want to go in the car later, we can take a look at where it was."

"No," he said. "I can see it in my head."

AT THE PLACE WHERE ROGERS' DRUGstore used to be, we paused and Jack said, "We brought in the autograph book."

"With my sister," I said.

"I think that one poor old janitor thought we were making fun," he said. "I felt so bad. I don't think he could write."

My younger sister, when we were very young, had received an autograph album for her birthday, and she wanted to get the autographs of some famous people, but we knew no famous people. So we walked to the place that was the next best thing: Main Street, where grown-up people worked.

My sister and Jack and I went to the gas station, we went to the insurance office, we went to Rogers' Drugstore, asking the men and women at work if they would sign the book. We didn't mean it as a joke—we wanted their autographs. Most were very nice about it, and I do think that Jack was remembering correctly. The janitor at the drugstore, as I recall the moment, was awkward and a little irritated at our request—he was a man who did not read or write, we were told later, but we didn't know it, the last thing we would have done was try to hurt his feelings. We were on Main Street, that's all. We were small and Main Street was big.

"The air-conditioning in that store was always so freezing," Jack said, tasting it. "And up by the front cash register, it smelled like bubble gum."

"THAT'S WHERE WE MET THE BEATLES."

He was gesturing toward a store that had in recent years sold linens.

"We took your dad's big blue Pontiac, with the white leather seats," I said.

It was a mid-January night in 1964; the first *Ed Sullivan Show* appearance had yet to occur. The Beatles, at that point, were still little more than a rumor, their voices coming out of our alarm-radios each morning, via WCOL, to help stir us to wakefulness. We'd seen their picture in *Time* magazine, we knew they were on their way to the United States, but there was nothing available to buy. No way, yet, for a sense of ownership.

We called Bexley Records every day. A wooden-walled and sparsely stocked storefront set back off Main Street, it was run by a couple of serious-faced young men who were folk music enthusiasts; a musician could purchase strings (either nylon or steel) for acoustic guitars, the latest Kingston Trio and Brothers Four albums were generally in stock, the sounds of Odetta or Joan Baez usually wafted through the store from the record player behind the counter.

The two proprietors of Bexley Records were becoming accustomed to our calls. Theirs was not a store that as a practice carried rock albums, but they had placed an order for *Meet the Beatles,* which was going to be the band's first American LP, and while they could not promise to save copies for us, they told us that if we would check in each day, they would let us know when their shipment arrived.

They were true to their word; on a snowy night Jack borrowed his father's hulking boat of a car and we drove to Main Street and there were the albums, with the black-and-white portrait of the four young men on the cover. Not even shrink-wrapped, as I recall, just the square cardboard album cover and, inside, a piece of round black vinyl that would make the world seem a little different, a little better. Two dollars and forty-seven cents an album, and we each bought one and hurried back to my house to listen.

"It really did feel sort of like meeting them," he said now. "All the other songs on the album that no one had heard yet."

"I Saw Her Standing There" and "I Want to Hold Your Hand," the two sides of the single, had been getting constant airplay on WCOL, but everything else on the album—"It Won't Be Long," "Don't Bother Me," "Not a Second Time," all the others—was unheard, fresh. When we played the album that first time it felt like receiving long-hoped-for phone calls from people we'd been wanting to know. The voices; the words. *Meet the Beatles.*

"When I finally did meet them, I don't think it was as good as that night," I said.

"That's what you said when you called to tell me," Jack said.

In later years, after I was on my own and working, I'd run into George Harrison and Ringo Starr on separate occasions—Harrison in a hotel in Jamaica, where he happened to be vacationing while I was covering a story, Starr on a trip he took that had him pass through Chicago. They were pleasant men, each in his own way. Yet there was no flighty thrill to the encounters, no dizzy sense of discovery—we were all older; asking the merchants on Main Street for their autographs had been a giddier experience than chatting with the two Beatles. The kick had been long before—on the winter night at Bexley Records—when we had gotten the album before anyone else in town.

"My dad asked us why we each had to have one," Jack said. "Why we couldn't share."

Jack's mom had been gone for less than a year that night he brought his album back to his house.

"We asked your dad if he wanted to listen," I said.

"He didn't," Jack said.

THE WORD WAS "LOOPY."

I don't know whether Jack came up with it, or Chuck. They had both been using it, in these months since he had become sick, to describe something that was happening to his cognitive skills.

Jack didn't mind the word; he said it himself to explain what was going on.

"I'm just feeling a little loopy all of a sudden," he said as we stood near the door of the store that in its recent incarnation was devoted to selling bedsheets, the linen store that once was Bexley Records. "Am I repeating things?"

He was, a little bit. We didn't know if it was the cancer, or the dreadful strength of the medicines intended to stop the cancer, but he would forget things, he would tell you something one minute and then, an hour later, he would tell it to you again. Janice had said there were days when he would ask her to do something for him three or four times, and it wasn't that he was haranguing her, it was that he wasn't certain he had asked her in the first place.

I told him that, yes, he had repeated some things this day, but that he and I both knew why he was doing it, and that he shouldn't worry unduly about it.

"It's not like I don't know when I'm getting loopy," he said to me. "But knowing it doesn't make it go away. It kind of frustrates me."

He and Janice had looked into the home-health-care situation. It wasn't time for nurses in their house yet, but they had made all the arrangements for when that time would come. The doctors were recommending a different medicine for the chemotherapy sessions. The first had not had the hoped-for effect.

AT THE BEXLEY PUBLIC LIBRARY HE FELL silent.

So much, I could tell, was flowing through him. The library—the soul of Main Street—was once dim and a little musty, now lustrous and determinedly airy: Recent reconstruction and redesign had rendered it a more physically welcoming place than when we were boys. At least that was the intention. The fact is, there was never a moment when that building was unwelcoming to us.

When we were children, the library was the place to sit on the floor on sultry summer afternoons as a librarian read to us as part of story-hour groups, while outside the stone walls the sun beat mercilessly on the Main Street sidewalks. When we were a little older, and had learned our way around the card catalogs, we would come here by ourselves in search of weekend reading: the Freddy the Pig books by Walter R. Brooks; later the Chip Hilton sports novels by Clair Bee (before we were quite ready to graduate to the slightly more advanced works of John R. Tunis); the Tom Swift Jr. books by Victor Appleton II. In high school our interests had shifted; we'd come to the library on autumn nights in the guise of doing research for term papers, but our real goal was to run into certain girls—to accidentally run into them, on purpose—and persuade them to leave with us. Sometimes we succeeded.

Now Jack stood in the foyer and, just as before, the sound of the library was a sound like no other: shoe bottoms clicking against the marble floor, the clicks, in the midst of quiet, the prevailing noise, if you can call a sound so lovely noise.

"The way those books would smell, back in the stacks . . . ," he said.

"The bubble gum at the drugstore, the books here—you're coming up with a lot of fragrances today," I said.

"They smelled like dust, and binding adhesive," Jack said.

He was tasting everything.

BROADWAY PLAYS WERE SOMETHING HE educated himself to love.

I'd never been to one—not in New York; not on Broadway.

"You've got to do it," he would tell me, over the years. "You can call these services, or go up to these booths on the day of the show, and you can get tickets at really good prices . . . Going to those shows have been some of the most incredible nights of my life."

He would go alone. He'd be in New York on business, he'd be staying in a hotel by himself, and he'd pick out a Broadway play or musical.

"You never ask yourself whether you belong there?" I'd ask him. That, probably, is why I never had gone—Broadway always seemed so far, in every way, from here. A ticket could get me in, but I knew it wouldn't make me feel like I fit in.

"You belong there the second the curtain goes up," he'd told me. "You belong there the moment the orchestra hits its first note."

Now we were in front of the Drexel Theater, Bexley's only movie house. In there, I'd always felt at home. In there the movies, when we first were going to movies, had mostly been second-run, and inside the Drexel, at the intersection of Drexel Avenue and Main Street, the panoramas—even the most gargantuan Hollywood panoramas—had seemed reassuringly and ultimately life-sized. We had known that, off in New York, sophisticated people were amusing themselves at big-ticket Broadway shows: *My Fair Lady, The King and I, West Side Story,* and afterward, in our imagining of their nights, they were stopping off at Sardi's, at Toots Shor's, places we would never see.

We would walk out of the Drexel—after our fifth or sixth viewing of Kirk Douglas in *20,000 Leagues Under the Sea*—and step right next door to the Feed Bag lunch counter for a milkshake or some french fries, and that was just about as worldly as we required matters to be, that was just about as much entertainment and post–theater cuisine as we could ever covet.

But Jack had become a man with an appreciation for Broadway shows, he had given himself that, and at the Drexel now he said: "I'm hoping that when I go to Sloan-Kettering I feel good enough to go to the theater that night." The competing forces: his life, and all its desires and drives, against that which was sapping his capacity to chase those desires.

Not knowing what else to say, I said, "There's no Main Street in Chicago. I used to think that was really strange. What kind of city doesn't have a Main Street?"

"Now you're the one who's loopy," he said.

WE KEPT SEEING THE PLACES THAT weren't there.

Wentz's Pharmacy was now something else. So was Evans and Schwartz Shoes. The Glass Bowl, and Willard's Family Restaurant, Paul's Food Shoppe and Seckels 5 & 10 . . . we walked by the buildings and it was like seeing ghosts. Except ghosts are gauzy, ghosts float in and out of your field of vision; the ghosts of Main Street were ghosts that had left calling cards, they were there and they weren't there. The buildings, most of them, remained, with new signs, new owners, new purposes. We saw what stood there now, and we saw what went before.

Jack was quiet again, as he'd been inside the public library. With anyone else I'd have pointed out what we were seeing—the Main Street that used to be, the Main Street that had replaced it—but Jack knew what I was seeing and feeling, no words were necessary. This had all been here, had thrived, and now: gone. All these stores, crucial to their community: supplanted. Expendable.

I didn't have to ask what he was thinking.

"I need to get something to eat," he finally said. "An ice cream soda or something."

We went into Graeter's Ice Cream Parlor—in the old Wentz's building—and Jack asked for some low-fat ice cream. He was told they didn't have any; would he like a soda made with regular ice cream?

"No, thanks," he said. Then, to me:

"Regular ice cream's no good for my cholesterol."

I had to stop myself from telling him it didn't matter. I had to stop myself from saying: At this point, enjoy whatever you want. An ice cream soda with real ice cream in it is no longer your concern; if you want a soda, have them make it with their best ice cream. The outcome of this woeful game will not depend in any way on your sticking with a prudent dessert.

But it was low-fat he wanted, and we walked down Main Street to Johnson's, where we had gone on summer afternoons for ice cream when we were boys. It was a Monday, though; Johnson's was closed on Mondays. We'd forgotten.

"What's that across the street?" Jack said.

I didn't see where he was looking.

"Right there," he said. "That Anthony-Thomas place. Maybe they have sodas."

It was a candy store. I'd never seen the building before. It had obviously, from the look of it, been there for years. But it was tucked into a little parking lot through which I had never walked.

"Let's cross and see if they'll make me a soda," he said.

"We can go over there, Jack," I said. "But they're not going to have sodas. They sell candy."

"Maybe I can get a soda there," he said.

Loopy, or tired, which I didn't know. But we crossed Main Street, and there was one young guy working in an otherwise empty store. There were glass cases filled with chocolates. Pieces of candy, boxes of candy. That's all.

The young man looked surprised to have customers. "May I help you pick something out?" he said.

Jack looked at the contents of the glass cases with some interest—he was, after all, the man who had just shipped five trucks of candy across the nation. He evaluated the goods, then said: "I'd like to get a chocolate soda, with low-fat ice cream."

The young man said, "We don't have ice cream here. We don't make sodas. Have you tried Johnson's?"

"It's closed," I said.

"It's Monday," Jack said.

We thanked him and left.

"It was worth the trip," I said to Jack. "We finally did the thing we thought we'd never be able to do."

"What's that?" he asked.

"We found a place we've never been," I said. "We went somewhere in town we'd never seen before."

"I'm going back in there," he said, turning around.

"Why?" I said. "He told you they didn't have ice cream."

"I feel sorry for him standing there with no customers," he said. "Let's give him a little business."

Seventeen

THE PHONE RANG LATER IN THE EVENING than was customary for him to call.

I was back in Chicago. He had been to New York, to Sloan-Kettering; the miracle he had been waiting for had not happened. The doctor there looked at his charts, went over the X-rays and CAT scans and blood values, spoke with Jack and Janice for a long time . . . and concurred with everything Jack had already been told by his other doctors. It was fine to try to change the mix in the chemotherapy, the doctor said—it couldn't hurt. But there was little reason for optimism.

These things, though, I had known about for a few days. Jack and I had discussed them at length.

So that couldn't be what he was calling about, later than he usually did.

I answered the phone and he said that he and Janice had been somewhere.

SOMETHING THAT HAPPENED ONCE WHEN he was living in Minnesota—something that he didn't consider any big deal:

I'd called his house in the Minneapolis suburbs to speak with him, and the phone had been answered by an unfamiliar voice. There was a radio blaring in the background, and for a moment I thought I might have dialed an incorrect number. The voice on the other end said a few labored words that I couldn't understand, and then stopped. The rock music from the radio continued.

I stayed on the line, concerned that something was wrong. Three or four minutes passed, and finally I heard another phone being picked up in the house. I heard Jack's voice say, "Hello?"

I said hello back, and I said I'd thought that a foreign-born person or a person who was unable to speak coherently had answered his phone; I asked if anything troubling was going on.

"Tell you in a minute," he'd said, and his voice went away and soon enough I could hear it in the room where the other phone was off the hook—the room from where I was hearing the rock music.

I heard Jack say: "I've got it, honey." Then: "It's all right. Thank you for answering."

He had hung up that phone, and when I heard him again he could talk without her listening.

"She's such a sweetheart," he said, and then explained.

He had been out of town on business, and when he

returned the girl—the girl who had answered the phone with such difficulty—had moved into his house.

Janice was working as a substitute schoolteacher, and at one school where she taught there were children who had severe disabilities. This girl was one of those children. She had been born with a grievous problem that had affected her both physically and mentally. It was not the kind of infirmity that surgery or medicine could fix.

When Jack had been out of town, the little girl had suffered another painful setback. There had been an accident in her house, and her mother was going to be unable to care for her or her brothers and sisters for a while. The girl—she was twelve—didn't really understand. One more rotten circumstance handed her by life.

Instead of letting the girl be turned over to a social service agency, Janice arranged through the school to take her in until her mother got better. Jack returned from the trip, and she was living in his house. It had been a family of three—Jack, Janice and Maren. Now there were four, at least for a while.

"It was our anniversary," he said. The day he got back from his trip was his and Janice's wedding anniversary.

"The four of us went out to a nice restaurant," he said. "I don't think she'd ever been anyplace like that in her life. She's just a wonderful child. She looked so beautiful and happy. It was like she was a princess in a castle."

I had asked him if even a part of him would have preferred to have had the wedding anniversary dinner be more private. Just him and Janice.

"Are you kidding?" he had said. "You should have seen her face, looking around the restaurant, and ordering from the menu. I don't think I've ever had a better anniversary dinner."

The girl would be going back home at some point. Maybe her life would be richer for having been treated with such kindness and care during her own family's anxious time; maybe she would never really understand.

And Jack didn't think what he and Janice had done was all that noteworthy a gesture. Someone had needed some help. That's all.

IT'S A QUALITY THAT CAME TO HIM EARLY. In school, some people who didn't know him well had a bantering nickname for him: "Class." It wasn't meant as a takeoff on "classy"—it referred to his perpetual willingness to assist with whatever our class (not one classroom—the entire class that would eventually graduate together) was in need of.

He ran for and was elected an officer of our senior class—it was one of the few times he volunteered to me the void his mother's death had left in his life ("Man," he had said to me after winning the election, "I wish she were here to see this. She'd be so proud"). When there were Thanksgiving canned-food drives for disadvantaged families in Columbus, he'd be the one who never needed to be reminded by C. W. Jones's incessant announcements on the school's public-address system—the first day of the drive he'd usually come into school lugging a bushel basket full of canned goods (and it couldn't have been free food from his dad's market—Mr. Roth was a purveyor only of fresh produce. Jack went over to the Kroger supermarket on Main Street by himself to buy the requested cans of groceries). It sort of evened out the ABCDJ cosmic account . . . Dan and I steal the chicken,

Jack shows up weighted down with food for the indigent.

In the high school fraternity to which we belonged, Jack was the pledgemaster senior year, and the freshmen probably had no idea of what a break that was for them. Pledgemasters, by definition, are often closet sadists, dreaming up new cruelties to inflict upon those who strive to become active members. With Jack that year, the pledges got a "master" to whom they could confide their problems, who would excuse them from doing tasks if they had a cold or a stomachache or were otherwise under the weather, who lobbied to convince the rest of the fraternity that mordacious pledging was a misguided tradition, one that reflected far more poorly on those who would carry it out than on their intended prey. Pledgemaster? In Jack, the freshmen had more of a union steward, standing up for their rights against entrenched management.

He simply could not abide meanness, all through his life. There was one boy at school who suffered from epilepsy. He was the kind of kid who found himself left out of just about every social and athletic activity. He would get spells sometimes—not seizures, but almost a trancelike state. Apparently it was the result of his medication.

One day, when he was beset by one of those spells, a bunch of guys gathered around him and started to torment him. They saw weakness and they pounced—insulting him, laughing at him, trying to confuse him, making an already distressing moment for him even more difficult to bear. Jack saw what was going on—the boy, in addition to being incapacitated, was outnumbered—and called Chuck and me over to help him break it up. We did; it never seemed to occur to the boy's persecutors that there was some other way to react to his defenselessness than to taunt him. Chuck and I walked him down to the nurse's office so that his mother and his doctor could be called, but it was Jack who noticed, it was Jack who would not allow what was happening to continue.

It was something that was built into him—not just a part of who he was, but the essence of who he was. In most people, you don't notice its absence: Indifference, after all, is usually not visible. When overriding mercy is present, though, it shines like a beacon; when it's there in a person, only then do you perceive how many others lack it, and what a rare and good thing it is.

NOW HE WAS CALLING ME AT A TIME OF night when he usually didn't call, and he said: "Janice took me to a support group tonight."

It was an organization she had found through the hospital, he said; the group, led by a counselor, consisted of cancer patients and their families, with the purpose of talking about their mutual experiences.

"Doesn't sound like your kind of thing," I said.

"Yeah, tell me about it," he said. "I had resisted going, but this has been very tough on her, and she said that if I didn't want to go for me, would I please give it a try for her."

So that is how they had spent their evening: in the company of strangers, all of them with one thing in common.

We grew up in the era before the phrase "support groups" became a part of the national idiom; we grew up during a time when, if tragedy befell you, you were pretty much on your own. Your family, whatever was left of it, was there; your friends were there. Maybe a member of the clergy. But mostly you were the one who had to deal with the devastation yourself.

Most young people never would have had occasion for a support group anyway; most were lucky enough to make it to adulthood with the basic framework of their lives intact. Jack wasn't so lucky; maybe when his mother died he could have used a support group, or maybe he would have rejected the idea even back then. But the choice was not his. Formal groups to offer counsel and condolent perspective to people with lives suddenly torn asunder were not in plentiful supply. We were Jack's support group then, and I suppose we didn't do the most expert job in the world. We didn't know how.

Now Janice had taken him to the support group meeting, and he had called to tell me something about it, but what he had to say was not what I was expecting.

HIS WHOLE LIFE, I NEVER SAW HIM PANIC. I'd seen him just about every way you can see another person. I'd seen him joyful, I'd seen him distraught, I'd seen him full of laughter, I'd seen him proud. I'd seen him angry, ambitious, exuberant, downcast, afraid, revved up, on top of the world, wounded, reflective, delighted, disconsolate, mirthful, aggrieved, resolute, contented. There was no way I hadn't seen him.

Except one.

I had never seen Jack panic.

There was a calmness at his center—there always had been. In the midst of everything, always, from the beginning to now, it was immutable. I never knew where it came from, but it was there.

I would say to Chuck, in these months since Janice found Jack on the floor: "Is he back from radiation?" Or "Is he back from chemo?" I'd ask it the way I used to ask "Is he back from lunch?" It had become that routine a question in these months. And Chuck would always tell me how Jack was doing on the day in question, and of all the ways he was doing, the constant was the lack of panic. The constant was the calm.

Even now.

SO HIS VOICE WAS STEADY AS HE TOLD ME about the support group meeting.

They had all sat around a room, he said—around the periphery, in the rough approximation of a circle.

The counselor had been in charge, but the idea was for each patient, and each spouse or family member, to have his or her say. The counselor was there mostly to keep things going—there was no specific agenda or timetable, the reason for being there was to talk about whatever came to mind.

Treatments, fears, surgeries, forays into experimental drug therapies—nothing was off-limits. All the things that the patients and their loved ones seldom said aloud anywhere else, they were encouraged to say here. To an audience of equals, an audience that would understand.

"Did you talk?" I asked him.

"A little bit," he said. "You don't have to, but you feel funny if you don't say anything."

I wasn't certain why he had placed the nighttime call to me—usually by this time of evening he was weary and ready for bed. If he hadn't done much talking at the support group meeting I didn't know why he wouldn't elect to wait until morning to fill me in.

Then he said it:

"Greene, I just felt so bad for them."

I thought I might have misheard.

"Bad for who?" I said.

"Those people in the group," he said.

"But aren't they going through the same thing you are?" I said.

"They're much worse off," he said. "They've all had it longer. Some of them have had these really serious operations—some of them had to be assisted into the room, and really seemed to be in a lot of pain. My heart was just breaking for them."

That is why he had called. Not because he was feeling sad for himself. Not because he saw his own future when he looked at them. Jack had cancer of the worst kind, he knew that death was on its way, yet all he could think about was the hurt being endured by the strangers with whom he had spent the evening.

"You see something like that, and you just wish there was something you could do to help," he said.

He said he was going to have trouble sleeping that night.

"I know I'll be lying awake thinking about them," he told me. "No one deserves to go through what they're going through. I couldn't find the right words to tell them that things might turn out all right."

THE NEXT DAY I WAS SITTING BY MYSELF in a restaurant in downtown Chicago. I was thinking about Jack, of course; I was making plans for my next trip to see him.

A family came in with a grown son who was disabled, and whose affliction caused him to involuntarily create a bit of a commotion. It must have been a neurological disorder; the sounds and jerky motions he was making were beyond his control, and some other diners in the white-tablecloth restaurant turned around in initial annoyance.

I saw a waitress, two waiters and a manager approach the family's table. The members of the restaurant staff had seen the disapproving reaction of the other diners; I was afraid they were going to ask the family to move to a more remote part of the room, or even to suggest that they might select a different restaurant. To do such a thing is a violation of the law, but it happens.

The waitress, in a bright and cheerful voice, said: "Hi, Ray!" She knelt next to where he was sitting.

"How you doing?" she said to him, smiling and looking into his eyes.

"We've missed you," one of the waiters said to him. "Where've you been?"

The manager said, "Welcome back, Ray! How's everything going?"

The young man was beaming. He'd been here before; they knew him. They welcomed him. They immediately put him at ease.

And his family was so pleased. I got the impression

that it was no coincidence they were repeat customers. There must have been many places where the reception for them was not so caring. There must have been many places where, because of the noise and tumult inherent in their son's condition, the stares were mostly cold.

Seeing the staff with the young man, the other customers in the restaurant broke into appreciative smiles. That's all it had taken—to see the family treated with such kindness and understanding was all it had taken to make the other diners regard the presence of the young man not as a minor inconvenience, but as a benefaction of sorts. A reminder of who all of us, at our most graceful, can be—on both sides of the table.

And maybe it was just the mood I was in, maybe it was because I was still marveling at what Jack had said to me the night before, but I thought: Jack, there's your legacy, right there. Something so simple. People being good to other people in trouble, people being good to other people who can do absolutely nothing for them. People noticing others' need for warmth at the exact moment when that need is at its most unguarded.

Maybe it was just the mood. Doubtless I'd get over it—the impassive causticity of city life would soon enough take over again. *I just felt so bad for them,* Jack had said the night before, seeing only the suffering around him, pushing his own troubles completely to the side. *We've missed you, Ray,* the restaurant staff, gathering about, said today, in that one instant filling the young man's face with joy, setting his family's hearts at peace. You're here, Jack. That's what I was thinking. You're not alone. The best of you is the best of us all. You're here.

Eighteen

He was upbeat and his voice was full of life when he gave me some unexpected news:

"I went to the fair—twice."

"You did?" I said. "Who'd you go with?"

"Janice and I went with some friends," he said. "Oh, man . . ."

"Did you feel old?" I said.

"You start feeling old at the fair when you're twenty," he said. "But it was great. It was . . . you know."

"You see the butter cow?" I asked.

"What do you think?" he said.

HIS TRIPS TO THE OHIO STATE FAIR came at around the time his doctors were suggesting he add something to his regimen: oxygen.

He detested the proposal—the idea of it, the logistics of it, what it represented.

The rest of us didn't quite understand why he resisted it so adamantly. His lungs were shutting down—gradually, yes, but there was no denying it. His doctors thought he wouldn't have to exert himself so much while breathing if he got a little help from a portable oxygen tank. Thin, clear tubes, they said, would be provided to let the oxygen flow into his nostrils. The objective was to make things less taxing for him.

With all the other ways his life had changed—the chemotherapy, the radiation, the bottles of pills that now lined the surface of his bedroom dresser—the oxygen seemed like just one more increment, and not that drastic a one. The doctors' intention was for it to help him out. To make him—in every sense—breathe easier.

It made him angry. It made him depressed. No wonder he sought out the fair.

HE STILL, SOMEWHERE IN HIS HOUSE, HAD the pencil portrait of himself he'd gotten on the midway of the fair when we were twelve or thirteen. I'd lost mine long ago, but Jack had kept his.

The Ohio State Fair—the sun-baked annual end-of-summer celebration of food, music, rides, farm animals, carny games and aimless all-day, all-evening wandering—was the state's biggest, longest party. At least it always seemed that way when we first knew it.

Held for two weeks on the sprawling fairgrounds up by the Ohio State University campus, it was the last blast of summer—the last good times before the end of vacation, before the unforgiving timetable of the real world kicked in again in the fall. Yes, it was corny—literally, the place was rife with corn; corn on the cob, corn dogs, corn on a stick—and you weren't supposed to take it seriously, even when you were a kid the idea was to approach it with a wink. Life itself might be austere and leaden and assiduously regulated. The fair was the antidote. The fair was a 360-acre grin.

You'd ride the bus to the Eleventh Avenue entrance to the fairgrounds, walk through the gaps in the towering letters of the O H I O gate, and immediately you'd be in a place you'd been forever, you'd know exactly where you were and who you were. You were a kid in the middle of the country at the end of another summer—no matter how many times you'd been there, you were always that kid at the Ohio State Fair. The farm families would be present, having driven in from every one of the state's eighty-eight counties to display their prize livestock and finest crops in competition; city people would arrive for the nighttime concerts in front of the grandstand; there would be harness racing and live radio broadcasts and, when you were young, nonstop flirting with people you'd never seen before and would never see again.

There was Bobo, the guy who sat in the dunk tank, yelling insults as you passed by until you were sufficiently worked up to hand over your money, take three baseballs, and throw them at him to try to drop him into the dirty water; there was the big cow carved lovingly out of pure and beautiful Ohio butter and kept in a refrigerated glass case for fairgoers to approach and pay annual homage; there were sketch artists on the midway to draw your portrait, maybe the first professional portrait of yourself you'd ever owned, and maybe the last.

It's the one Jack still had; his fairgrounds pencil portrait still was in his home. And of course this year he would want to go to the fair. It shouldn't have surprised any of us. His doctors were telling him he should consider an oxygen tank, and this made him furious, he knew they were right but the fact that they were right upset him in ways the rest of us could only begin to grasp. Of course he wanted to go to the fair: the last breath of summer.

I THINK HE FIRST FOUND HIMSELF—HIS confident and self-assured self—at the Ohio State Fair. I'd never mentioned it to him; if I had, chances are he'd have told me I was wrong. But I knew him pretty well. I'd seen it.

One summer our friend Pongi's dad had the soda-pop concession contract in the grandstand. For all the outdoor shows—the horse races, the rock concerts, the tractor pulls, the finals of the livestock competitions—Pongi's dad got to sell RC Cola, and the other flavors his bottling plant produced, out of paper cups. He hired hawkers to roam through the stands selling.

It was a salary-plus-commission job for the hawkers; they got a little money just for doing it, but their payday really depended on how many cups of pop they sold. As I recall, they had to go underneath the stands and pay in advance for each tray of RC they took out; when they came back for refills they were allowed to keep a percentage of the receipts for each tray.

Jack was hired. Like most kids, he'd always been a little shy in social situations, reluctant to assert himself with people he didn't know. And for most people, that reticence eventually, to one degree or another, tends to go away over time. With Jack I saw it happen.

It was that summer—the summer he hawked RC in the grandstand. For the first time, he wasn't working for his father in the market; for the first time, the faces he saw every day were new faces. People moved in and out of the grandstand—sometimes there were three or four grandstand shows of different varieties each day. Jack and his fellow hawkers stayed on; the people in the seats changed.

He would come home at night—meeting the rest of us back in Bexley at the end of the evening—and he was suddenly sure of himself. He'd smell of cola—the RC splashed around a lot when the cups were being filled beneath the stands, by the end of a shift his forearms would be sticky from the foamy, sugary liquid and no matter how hard he scrubbed, the aroma of RC would remain. But he'd have talked, however briefly, to hundreds and hundreds of people each day; he'd have made eye contact with dozens of girls, been openly flattered by the things some of them had said to him, he was feeling like his own person, a person in the wider world. And this was in a context other than ABCDJ; Allen, Chuck, Dan and I didn't work at the fair. This was brand-new for Jack; this was something he was trying on his own.

We'd see him at midnights that fair summer, after his

evenings in the grandstand, and he was changed, if ever so slightly. He'd taken his first steps beyond everything he'd always known; he'd taken his first steps beyond his house, and his school, and us. Yet the stratosphere he'd put one foot into was the friendliest stratosphere a guy could traverse. If you've got to begin your lifelong journey into that more expansive universe sometime, you can consider yourself lucky if those first transitional steps take you straight through the Ohio Fairgrounds.

"WE SAW THE RASCALS, THE GRASS ROOTS and the Turtles," he was telling me now.

That's one of the things he'd done at the fair at the end of this present and quite separate summer: He'd gone to an oldies concert with Janice and some other couples they knew.

"In the grandstand?" I asked. I pictured him sitting in the same seats through which he used to roam with his soda-pop tray fastened to him by a strap around his neck.

"The *grandstand*?" he asked, as if I couldn't be more behind the times for not knowing. "The grandstand has been gone for years."

A new concert venue had been constructed; he said the show had been terrific, he had let the music fill him up, he didn't get tired that night and he didn't ache and he didn't think about the uninvited new life to which he'd be returning at evening's end.

"You eat there?" I asked.

"Of course," he said. "Ribs. Corn."

He said he'd stopped to pause for a few moments by the butter cow; you couldn't not.

And he answered my next question before I asked it.

"I didn't see Bobo," he said.

New generations of Bobos had come and gone in the years since we were first at the fair; they were always called Bobo, they were always yelling insults from their dunk tanks, but the Bobos themselves changed. Pretty frequently, I would imagine; being Bobo would not seem to be a long-term career move.

"Did you look for him?" I asked.

"Well, I wasn't going to walk all over the fairgrounds looking, but I looked everywhere we walked, and I didn't see him," Jack said.

"You think they got rid of him?" I said. "You think the Bobo tank's not there anymore?"

"Come on," Jack said. "The fair with no Bobo?"

He'd enjoyed himself so much that he'd gone back for a second night. He couldn't get enough of it.

I wasn't sure whether to ask, but I did.

"You take the oxygen?" I said.

"To the fair?" he said, meaning: No.

NOT ALL ATTEMPTS TO DIVERT HIMSELF were as successful.

Chuck and Joyce went on a summer vacation trip to Colorado, and invited Jack and Janice to join them. The idea was to do nothing but relax—dinners, music, walks on mountain paths if Jack was up to it.

But Jack, on that trip, wasn't up to much. Probably the altitude had something to do with it—they had all talked about it before they flew out there, they had made some inquiries about whether being up in the mountains would be too much of a burden on Jack's respiratory system, and they had been told it was safe for him to give the vacation a try, if he wanted to. He wanted to.

Once he got there, though, everyone knew it wasn't working. "Every night we'll go somewhere or other," Chuck told me when I called out there and Jack was sleeping and couldn't overhear. "And he'll really want to come along. We'll get to wherever we're going and we'll be there for a few minutes and he'll start getting this look on his face. And after a while he'll say, 'I have to get back.'"

He wasn't taking walks, Chuck said.

"He just can't."

The fair was one thing; Colorado was harder. Probably it was the altitude. But maybe he sensed that he was too far from home.

THE SPIELBERG EFFECT, HE ASSURED ME, was still in full force.

"I know I look different because I've lost weight, but it's happened a couple of times when I've been out," he said.

He first noticed it years before, when he grew a small beard, salt-and-pepper gray. When he'd go out somewhere wearing a baseball cap—and especially when he wore wire-rimmed eyeglasses—he would get stares.

"At first I couldn't understand it," he told me back then. "I was a little self-conscious—I'd go places and people would be looking at me. I thought maybe I had food on my face or something."

Then, once, in a restaurant in New York—a very popular

place, difficult to get reservations—he was in the foyer area, waiting to be seated along with many other would-be diners, and was settling in for a long delay. The place was packed.

But the maître d' glanced at him, walked right over to where he and his business associate were standing, and said quietly: "We have your table, Mr. Spielberg."

He was confused; his business associate's name wasn't Spielberg, his own name certainly wasn't Spielberg—what was this about? He didn't argue. He thought it was a mix-up that had worked out in his favor.

After that, it kept occurring. Usually it was just the stares; usually it was out of town, not in Columbus. He finally figured out what was happening when someone approached him and said: "I know you must get this all the time, but I love your movies."

Jack always had a gentle, wise-without-being-aloof face. And in a baseball cap, with the little beard and glasses, he looked very much like Steven Spielberg. Specifically, Spielberg in those on-the-set production stills that show him directing his movies.

I'd never thought about it; once Jack pointed it out to me, I could see the resemblance, but it never would have occurred to me otherwise. What was interesting to me was that a movie director was so recognizable that the public would pick him out of a crowd—or, more to the point, pick Jack out of a crowd. You think of movie actors getting that kind of response, not directors.

It continued to happen. Jack without the baseball cap bore little similarity to Spielberg. No one would ever make the mistake. But with the cap . . .

He was checking into a hotel on the East Coast once and the entire way up to his room the bellman was talking to him about "his" movies. Jack was amused by it—he had been through it enough times, it had ceased to surprise him—and when he got to the room he gave the bellman a tip (I didn't ask him if he gave him a Spielberg-sized tip) and then shook his hand. He said, "I know what you're thinking. But my name's Jack Roth."

At which the bellman gave a conspiratorial cock of the head and said: "Don't worry, Mr. Spielberg. Your secret's safe with me."

In Colorado, he said, he still had gotten the Spielberg stares when he was out. But he wasn't going out that much, even when he returned to Columbus.

"WHAT WERE WE DOING IN THE DIARY TOday?"

I'd kept a diary in 1964. I'd been attending a convention of high school journalism students from around the state of Ohio, and one of the guest speakers, a teacher, said that the best way to make oneself a good reporter was to keep a daily journal. The teacher said that the discipline of making yourself write down exactly what happened to you every day, even when you didn't feel like writing, was good training.

So, for that one year, I did it. Jack knew I still had it—the original diary, written in pen and in pencil day by day, in a spiral-bound calendar-journal that an insurance company had sent to my dad as a promotion at the end of the previous year.

"What were we doing in the diary today?" he asked me one summer night, as I was getting ready to come to Columbus to see him again.

He'd been asking the question more and more. He was trying to get something back—a time, a feeling. Maybe it was the same reason he'd gone up to the fair twice.

So I pulled the old diary—with its imitation-black-leather cover advertising the Archer, Meek, Weiler Insurance Agency—from the drawer where I kept it, and I turned to the day, this day, but in 1964.

"Dennis MacNeil came to my house with a Vespa motorbike he'd just bought," I said to Jack. "He rode me around town on the back of it for a long time. Then he let me drive it myself."

"Am I in that day?" Jack asked.

"You are," I said, looking at the diary. "My little brother Timmy asked you and me if he could go out to dinner with us. We said yes. We took him to Howard Johnson's at Broad and James. We paid our bill and you and I each bought a pack of Tiparillos—those little cigars with the plastic tips."

"Are those the ones Ernie Kovacs's wife used to sing about in the TV commercials?" Jack asked.

"Edie Adams?" I said. "I'm pretty sure she sang the commercials for Muriel cigars—'Why don't you light one up and smoke it some time?'"

"Then what was Tiparillos?" Jack said.

"You know," I said. "The woman in the commercial played sort of a cigarette girl, like in a nightclub."

"Carrying the tray of cigarettes and cigars around in front of her?" he said.

"Right," I said. "Like you with the RC at the fair."

"*'Cigars, cigarettes, Tiparillos?'*" Jack said, lightness in his voice, remembering now.

"Yeah," I said. "Anyway, we bought Tiparillos after dinner at HoJo's."

"Did I smoke one?" he asked.

"Let me look," I said. "It says that I did, but you didn't."

"I didn't think so," he said.

"I smoked one," I said, looking at what I'd written at the end of that evening in 1964. "We threw a football around in my front yard. We each had the Tiparillos in our mouths—mine lit, yours unlit. We let Timmy play football with us."

"What else happened that day?" he asked.

I looked at the diary entry.

"Nothing," I said. "That was it."

There was silence on his end of the phone. Then he said:

"Sounds like a pretty great day."

RIGHT BEFORE I WENT BACK TO COLUMbus he said Lazarus was closing.

The department store, at the corner of Town and High, had been a part of downtown Columbus for more than 150 years. Now it was the store's last day.

"They announced last year that it was going to close, but it didn't seem real," he said. "But this is it. They're locking the doors for good."

We'd spent so many hours there—first with our mothers, shopping for back-to-school clothes; then by ourselves when we were old enough; later as adults, even after the store had begun its slow decline as more and more people chose to shop at outlying malls. Jack had worked there one Christmas season. The Chintz Room, where ladies would gather for lunch, dressed up for a day downtown, more than a few of them wearing white gloves; the spectacular Christmas lights, and the Santa Claus parade to kick off the holiday shopping season; the vast book department in the years before there was any such thing as Borders or Barnes & Noble (or, at the other end of the spectrum, My Back Pages); the display windows put together as lovingly and as artfully as something in a gallery in London or Paris, or so it looked to our Midwestern eyes . . .

Lazarus was as much a part of downtown as . . .

Well, if someone had said that the Ohio Statehouse was going out of business, that couldn't have felt any more wrong than this.

"You're not going down there today, are you?" I asked him.

"I was thinking about it," he said. "But I don't know."

"There'll probably be crowds for the last day," I said. "I don't think you need to fight your way through that."

I told him when I'd be coming in.

"Lazarus isn't supposed to die," he said.

Nineteen

ALLEN WAS AT THE BAR WHEN I GOT there, smoking and looking like he was waiting for a jury to come in, which in a sense he was.

I had called him in Canton to suggest we both spend some time with Jack; he hadn't seen him in a while. Allen had been due at a fund-raising event—something to do with the top echelons of Ohio politics and law—but he changed things around and here he was at the Top. His wife was with him; they both had cocktails in front of them and he motioned for the bartender to bring me one before I even could sit down.

"Bobby," he said, in a deep-as-the-briny-sea, comically accented drawl; he's been using that word, and that exaggerated voice, to greet me since we were kids. He'd gotten it from some black-and-white-era television series long forgotten—at one point it had been a joke whose genesis we both instantly understood, but our memory of exactly where it had come from had faded over the years. The greeting had lasted.

"Alby," I said in the same voice, the word and the inflection the products of that same lost-in-the-mists-of-memory TV show. It must have made us laugh, the "Bobby" and the "Alby," once upon a time.

"So," he said, looking at his watch.

"They're going to meet us here," I said.

When Allen gets nervous you can see it. His body coils up. I've never observed him in a courtroom, even though practicing his profession in court has been the dominant fact of his adult life.

But this was what it has to look like—this taciturn, quietly wound-up demeanor is what opposing counsel must see on decision days before the bench. When something that counts is on the line.

"Come on, sit down," he said, an edge to his tone. "You going to stand there?"

Impatient. Irritable. Waiting for the jury to return. He kept glancing toward the front door, waiting for Jack. One look would tell Allen volumes. He'd know within a second what the odds were for a favorable verdict.

WE TALKED ABOUT THINGS THAT HAD NO importance to us, just to pass the minutes; he tapped his pack of cigarettes on the surface of the bar.

His wife asked me how my flight had been and when I had gotten in; I asked her questions neither of us cared about. We waited.

Then the door from the parking lot opened and Jack

and Janice entered, and Jack flashed us that chipped-tooth grin.

"Oh, Jesus," Allen said under his breath, even as he was allowing a broad, too-vigorous smile to cross his face, even as he rose with his arms extended. Hiding his cards. Not letting Jack see what he was thinking.

"I'm glad you finally got here," Allen called to Jack. "Bob's been sitting here telling us the dumbest stories you could ever hear. You're lucky you missed it."

Jack had regressed even in the time I had been away—I could see that. He was much thinner, there was a hollow look in his eyes and an overall pallidity about him. Still, I had seen him more recently than Allen, and I suppose I had been hoping that Allen, encountering Jack for the first time in months, would be able to reassure me that things weren't that bad.

"Come on, let's get to the table," Allen said, and he put a hand on Jack's back. He was smiling ardently for Jack's benefit, and he turned briefly to look at me and even though he maintained the smile I could see that he was wanting to weep.

WHEN WE WERE IN OUR EARLY TWENTIES—during that period when Jack was teaching in the suburbs north of Chicago, when I was starting as a reporter at the *Sun-Times*—there was a Friday night when he rode the train downtown to meet me at a bar called Riccardo's, near the newspaper building.

It was a famous hangout for reporters; I had only recently worked up the nerve to go in there without thinking I'd be asked to show a passport and then be summarily turned away. Mike Royko, then still in his thirties, would be at the bar, and Bill Mauldin, after he'd drawn his editorial cartoon for the day; Mauldin's friend and fellow cartoonist John Fischetti of the *Chicago Daily News* would be on hand, as would my personal newspapering hero, sportswriter Jack Griffin of the *Sun-Times,* in his crewcut and his trench coat at the end of the bar, seeming more like Bogart than Bogart himself. Peter Lisagor, when he was in town from Washington, would come in, and this was a place quite far, in every way, from everything I'd known back in Ohio.

So when Jack joined me that Friday night, we kept sneaking little looks over at each other, translating to: "Can you believe this?" Translating to: "Are we really here?" We didn't say it—at the start of your twenties you seldom express out loud your wonderment at where life has suddenly taken you, there's something inside that makes you think you should pretend this is not all that big. But we knew; we looked at each other, and we knew.

Not so very long before we had been at the Toddle House on Main Street; not so very long before we had been purchasing tickets to see *Goldfinger* in its first run at the Loews Ohio Theater, just down the block from the then-thriving Lazarus. That was where we were from, and now we were here, and this felt more exotic than anything in *Goldfinger,* our being here seemed like something out on the far reaches of imaginability.

The evening grew raucous inside Riccardo's, and the drinks kept coming, and Jack was talking to newspaper people and television reporters and they were treating him just fine, he might as well have stepped into *The Front Page,* the Ben Hecht–Charles MacArthur play we'd both checked out of the Bexley Public Library not so many years before. In Riccardo's he had raised his voice so that I could hear him and he said: "Let's go to Las Vegas."

"When?" I said loudly back, in the midst of the din.

"Now," he said. "Tonight."

Neither of us had ever been there. In the jumbled dissonance of the bar, with the drinks doing their job on us, it seemed like a reasonable idea: Go out to O'Hare, buy tickets to Las Vegas, fly there without suitcases or changes of clothes, and . . .

"What do you want to do there?" I'd asked him.

"Whatever we end up doing," he said. "We can come back Sunday night and be at work on Monday."

And, as unlike him as this was, we almost did it. Being in Riccardo's, in Chicago at night, in the company of all those people, had intoxicated him in ways separate from anything having to do with the cocktails. We were different than we'd been back home—that's what it felt like. We were in Chicago now, and if that was possible for us then anything was possible—if we could be here, then we could be in Las Vegas tonight, too. Our old world wasn't our world anymore.

Or so it seemed until we left Riccardo's and walked out onto lower Rush Street. In the night air, without the noise from the barroom, there was a sense of deflation, or at least of reality, and as we climbed the stairs up to Michigan Avenue to get a cab to the airport, Jack said:

"What time do you think we'd get to Las Vegas?"

"Really late," I'd said. "But we'll pick up some time zones. And it never closes."

"Where would we stay?" he said.

"I don't know," I'd said. "One of those hotels."

"Do you have enough money with you?" he'd said.

"My checkbook's at home," I said.

We both knew by the time our feet hit the sidewalk in front of the Wrigley Building that we weren't going anywhere. The action inside Riccardo's may have fooled us into thinking we had turned into people who could fly to Vegas on a whim, who could step up to the craps tables (even though we had no idea how to play) and stay out until the sun rose above the desert and then start all over again . . .

"How much do you want to do this?" he'd asked.

"It was your idea," I'd said.

"I know," he said. "But it's been a long week. I'm kind of tired."

I knew.

"We can do it some weekend when we have time to plan," I'd said. "When we can pack bags and make reservations."

"I think that makes more sense," he'd said. "Don't you?"

So we'd gone down the street and gotten a cheeseburger somewhere; just us being in Chicago felt implausible enough, we didn't need Las Vegas to make us realize we were fish out of water, we recognized that quite pointedly right where we were. We never did go to Las Vegas together; I think we probably knew that evening that we never might. It didn't matter. Back home, we'd never even have gotten this far—we never would have entertained the idea of flying off to Vegas.

And now we were back home once again, at the Top once again, and Allen, a man who himself had seen the lights of Las Vegas many times, called for a new round of drinks for everyone, and in his eyes I saw something close to anguish for the friend he loved.

CHUCK AND DAN WEREN'T HERE TONIGHT. Chuck and Joyce had gone out of town; Dan and his wife were scheduled to take a vacation cruise leaving out of Fort Lauderdale. (Severe hurricanes in the Caribbean were bearing in on south Florida; Floridians were beating a retreat north on jammed interstates, were trying to get on overbooked flights to anywhere they could—all that mattered was that they flee the oncoming storms. I had said to Dan: "You're actually going to fly to Florida so you can cruise in the Caribbean?" He'd said: "Oh, yeah." I'd said: "Have you watched the news this week?" He'd said, in the staunchly pedantic manner of Woodrow Wilson addressing the League of Nations: "Go to Princess dot com," and although as usual with him I had no idea how that applied to what we had been discussing, at this point I knew further colloquy with Dan on the subject would be fruitless.)

So Allen and his wife, and Jack and Janice, and a friend and business associate of Jack's—a guy named Joe who had flown in from overseas—were at the table. This kind of thing was happening quite a bit: people who were in business with Jack, or who had known him as friends over the years, were finding reasons to come to Columbus. There was really only one reason: Jack. I think he knew it.

He was trying so hard tonight. He had told me that his ears had become increasingly sensitive to loud noises—there had been some concerts in Colorado that he'd had to leave, he said, because it had been uncomfortable for him to sit in the midst of highly amplified music. I could tell that, in the restaurant, every time the sound level rose it was getting to him. Once or twice I saw him wince.

But he wanted to be here. "Joe, tell them about . . . ," he said to his friend, the one we didn't know, and it wasn't Joe's story that Jack wanted Allen and me to hear, what he wanted was for his friend to be our friend, too. This, all of it, was important to him.

Someone we'd known and liked growing up came by the table—he was a podiatrist now, but that's not who he was to us, to us he was the guy in whose oddly shaped little driveway on Remington we'd shoot baskets—and he was very good about not directing the conversation to Jack's health, he was very smart about hitting just the right tone and telling a funny story and making it easy for all of us, and especially Jack, to laugh. But all through the evening the laughter at the table seemed like laughter in response to a carefully constructed screenplay—we were laughing in the right places, sometimes a little too heartily, we were laughing because we knew we were supposed to in defiance of the darkness before us. Back in April, when we'd had that first ABCDJ dinner, the laughter had been like a conferment, hopeful and precious. This was different. We were many months down the line, we knew more about where all this was going, and we laughed on cue, from somewhere in the throat, well north of the heart.

ALLEN WAS INTO THE POWER YEARS—those years to which most men in business seem to aspire all through their careers, and only some manage to reach.

The power years, or so I've observed, usually kick in—if they're ever going to kick in—when a man is in his forties or fifties. It's not just a matter of being successful, although success is a prerequisite for entering the power years. And it's not something that can be given to a person, like a promotion or a raise.

In the power years, there's an aura around a man—he's in charge, not just of other people but of himself. It's more than being a boss—you can be a boss in your twenties or thirties. The power years are, if not a state of mind, then something very close to it. In the power years, you report to no one but you—not on an organization chart, not just in a boardroom, but in your head. You see this all the time in prosperous and supremely self-confident men in their forties and fifties. Country-club grill rooms are the natural settings.

I never saw Jack covet the power years. It's not that he didn't make it—it's that it never occurred to him to desire it. Yes, he wanted to run his own operation and make enough money to do a good job of supporting Janice and Maren, but to be one of those guys—grill-room guys, power-years guys—had no appeal to him.

When I'd asked him once if he believed that a man's forties were when he first became eligible to enter the power years, he had thought for a second and then said:

"I don't think the forties are the power years. I think they're the fool-you years."

"What do you mean?" I'd asked.

"Well, you see these guys in their forties, the ones who build the biggest houses and get their pictures in the paper with their wives at black-tie fund-raisers," he said. "They have this look about them, you know? Like stuffed cats. Very satisfied.

"It's like they know they've achieved everything there is to achieve. They're on cruise control."

"So why are those the fool-you years?" I'd asked.

"Because it'll fool you," he'd said. "You become one of those guys, thinking it's smooth sailing from then on out, and those are the guys for whom things go wrong. The power turns out to be an illusion. They get fooled by it."

"Not all of them," I'd said. "Some of them go sailing right on."

"I bet you the ones who stay at the top are the ones who never believed in power years anyway," he'd said. "Just because they look like they have no worries doesn't mean they don't. The guys with more than a few worries in their forties and fifties are the guys who are going to be all right. The ones who assume they've won the championship of life—those are the ones who are going to get fooled."

He'd said this to me before life fooled him anyway. And he'd never assumed anything.

AT THE TABLE, THE LAUGHS WERE STILL screenplay laughs—everyone was trying too hard. You could hear it. Jack knew it.

At one point, though, he leaned over to me, a true grin on his face. He was thinking about something that made him want to laugh for real.

"You know when we were talking about the Forty Winks?" he said.

That's what he was thinking about—the old alleged call-girl motel down the street.

"I told Chuck that we were talking about it," Jack said,

"and he claimed never to have heard about it. Can you believe that Chuck's never heard of 'Change for a penny'?"

"Chuck's never heard of 'The Star-Spangled Banner,'" I said, and Jack lit up, he laughed out loud, and for a moment—a fool-you moment—we were somewhere else.

WE WANTED DESSERT, AND EVERYONE WAS tired of sitting, so we took the one-minute drive to Johnson's.

Full-bore autumn was on its way, but this was one of those nights that had one foot in summer. We'd been having milkshakes and sundaes outside at Johnson's our whole lives; the ice cream stand came into being right around the time we were born and here it still was, with picnic-style tables in front as if the view was of a rolling-to-the-far-horizon verdant meadow instead of the traffic on East Main Street.

So many lazy nights here, so many lingering conversations. No one ever rushed you at Johnson's. For the price of a cone you could sit with your friends and stay as long as you pleased.

"I still think you're wrong," Jack said.

"I'm not wrong," I said. "You didn't used to be able to go inside and eat. There was no inside. The only way to order your ice cream was through that little window cut into the front of the store."

"That's just not right," Jack said. "There was always an inside—they just made it nicer a few years ago. But you could always eat in there."

"No," I said. "There was nothing in there but the people scooping."

"Then where would you see the sign with all the flavors of ice cream?" he said. "The sign was inside."

He might have had a point.

But arguing about the history of Johnson's was safer than saying what was on our minds.

Allen wasn't talking at all. He was looking toward Jack with a remote expression on his face, sitting at a picnic table and smoking. I said to him, "I think, if you ask, they may have a cigarette-flavored ice cream." He mouthed me a common curse.

"Or a tobacco float," I said.

It felt like summer.

But it wasn't.

Allen stood and said he and his wife had to be taking off. He placed a hand on Jack's shoulder and said, "I'll see you," having no idea if it was true.

Jack nodded that nod.

So many nights, right here. A car pulled up; the person in the front passenger seat got out to buy ice cream for the driver and for the people in the back, and as he opened the car door I could hear the music, an old number from the *Hard Day's Night* soundtrack. John Lennon always did have that knack: to sing songs with words of absolute hurt, yet to make the songs seem peppy and bright because of the quick insistence of the melody. I don't know how he pulled it off: singing the saddest songs while keeping a rough-edged smile in his voice. Allen took one more look toward Jack, then turned to leave, Lennon's voice in the background singing "I'll Cry Instead."

Twenty

LESS ACCOMPLISHED SINGERS THAN MR. Lennon weren't quite as adept at masking their emotions. I can testify to that. I was one of them.

Beginning in 1992 and for the next ten years, in about as unlikely and unanticipated an alignment of the planets as I ever hope to encounter, I spent my summer weekends touring the country as a backup singer and two-songs-a-show lead vocalist with Jan and Dean, the surf duo from southern California whose mid-sixties hits included the number one "Surf City," "Dead Man's Curve," and "Little Old Lady from Pasadena." As with just about everything else good that ever happened in my life, Jack had something to do with how it came to pass. Not that either of us could have known it at the time.

On an April Saturday in 1964, Jack, Dan and I had gone downtown to Lazarus to buy records in the department store's fifth-floor music department. I had bought two singles—"I Am the Greatest," by Cassius Clay, and "The New Girl in School," by Jan and Dean. I'd written down that fact inside the insurance-company diary. Three decades later, in a piece of writing based on the diary, I referred to what happened that day, including the purchase of the Jan and Dean record. (Details were the strong suit of the diary; the teacher who had recommended keeping a diary had stressed that noticing tiny things was key.)

A musician named Gary Griffin, the keyboard player for Jan and Dean—who were still, after all those years, out touring—read what I wrote about the diary, told Dean about it, and the band invited me to join them on the road. I flew with them to a show in Kansas City, we all got along quite well, they invited me to sing on one song, one thing led to another, and before long I'd bought a guitar and was joining them on a regular basis. They played everywhere from football stadiums to minor-league baseball parks to corporate conventions, and one night the tour took us to Minneapolis for an event in a hotel ballroom.

Jack and Janice were still living in Minnesota at the time, and I asked them if they'd like to come to the show. I introduced them to the band—to see Jack and Dean Torrence sitting around before the show having a beer and talking warmed me in a way I couldn't have predicted. We'd cruised the streets of Bexley with Jan and Dean blasting out of our car radios, the world had grown more complicated over the years both for the famous singers and for the kids who used to listen to them, and somehow here we were. If Jack and Dan and I hadn't gone to Lazarus that day, and if I'd bought some other record—if I hadn't kept the diary—we would not be here this night. But all that did happen, long ago, and here were Jack Roth and Dean Torrence, absorbed in conversation together. Life, when you let it, can thrill you.

At the concert that night the audience was invited to dance in an area just below the stage. Most of the show was fast songs—we did a lot of Beach Boys music, "Little Deuce Coupe," "I Get Around," "Surfin' U.S.A."—but at one point Jan and Dean asked if the people in the crowd felt like slow-dancing.

They did. We went into "Surfer Girl," which had helped to form the seamless soundtrack of our Ohio summer in 1963, and the couples in Minnesota took to the dance floor and I saw that Jack and Janice were among them.

We sang, and that beautiful Brian Wilson song was just as haunting and just as heartbreaking as it had been so many years before; it still had the power to put you right in a certain place and a certain time. I sang and I looked at Jack, gazing into Janice's eyes as he held her . . .

And it was too much, I had to look away. It was too intimate—I don't know exactly why, but from the stage I saw my oldest friend, adoring his wife, and here I was, helping to provide the music, and the loveliness of it was almost painful, I felt I was intruding on something I shouldn't see. Onstage we sang: *Little surfer, little one, made my heart come all undone . . .*

Jack and Janice locked their eyes on each other, they were a man and wife deep into their married life, in love and holding on to each other and moving in slow circles on the dance floor, and I didn't feel I had the right to be witnessing this, it all felt too close. I knew I wouldn't be able to explain it to Jack—I knew I wouldn't even want to try.

How did we get here? In the Minnesota night the best friend I ever had embraced his wife and danced slowly with her to the song I was singing, the song we all had danced to when we were so young. *Do you love me, do you, surfer girl?* we sang, and Janice leaned up to kiss Jack on the cheek, and how did we get here, how do such things come true?

"HELP ME CARRY SOME BOXES FROM THE car."

Two days after our ice cream night at Johnson's, Jack and I were at his house, and he was asking me to give him a hand bringing some things inside.

"This is your car?" I said.

"Yeah, of course," he said. "Do you think I'm going to open the trunk of someone else's car?"

It wasn't that. It was that I had no idea what kind of car Jack drove. Even though I'd ridden in this one in the months since I'd been coming to see him, I couldn't have picked the car out of an automotive lineup. It was just one of those cars that people buy or lease—one of dozens of kinds of cars that all look pretty much the same.

"Did you ever think it would be possible for us not to know what kind of cars each other drove?" I asked.

"Well, the answer's easy with you," he said.

I didn't drive a car these days; living in the city, in Chicago, I'd decided many years before just to get in a cab if I had to go somewhere. Keeping a car in the city didn't make any sense. Still, I meant the question I'd asked Jack.

"Think about it," I said. "Could you have seen a day when I didn't know what your car was?"

"Put it this way, it's not a 409," he said.

Meaning: Now that no one writes songs about cars—now that cars, in the main, are transportation and nothing more—why should anyone much care what anyone else is driving? Jan and Dean had been a big part of the legend making—Jan and Dean, and the Beach Boys, and the Rip Chords: all of them convincing us that fuel-injected Sting Rays and Three Window Coupes and 442s and Jags were the ticket to ultimate happiness, that *buddy, gonna shut you down* was the mantra of American manliness, and that anything that mattered in life was most likely going to take place, or at least begin, behind the wheel of your car. To quote from car-radio scripture: It happened on the strip where the road is wide.

"Was it Doug Dauber who had the 409?" Jack asked.

"He had one of those cars," I said. "One of those car-song cars."

"Can you believe guys used to ask other guys what they had under their hood?" Jack said.

"I think the safe answer, if you didn't know or didn't care, was 'a three-twenty-seven,'" I said.

"Whatever that meant," Jack said.

We started lugging the packages from the trunk of his car—whatever kind of car it was.

"I think I lost track of the cars you drove by the time you were in your thirties," I said.

"You didn't miss much," he said, slamming the trunk.

THERE WAS ONE CAR RIDE THAT, IF IT HAD ended differently, would have meant that Jack and I would not be lugging boxes on this day.

One winter Saturday after we'd first gotten our drivers'

licenses, Jack, Dan, Chuck and I got into Pongi's Chevy for a ride over to Dayton.

We'd just wanted to be somewhere else for the day. Jack knew a girl over there—her name was Joyce Burick—and we went to her house, then cruised around Dayton.

After dark, we started the drive back to Columbus. It began to snow, and then the snow picked up, and the road became covered with ice. This was before the interstate in that part of Ohio, this was a highway with no median island—the lanes of traffic came right at each other.

Pongi hit a patch of ice, and we started to swerve. He was fighting the steering wheel, pumping the brake, but nothing was working. We swerved once into the lane of oncoming traffic; Pongi wrestled the car back into our own lane. The next swerve was more severe. He was helpless to stop the skidding—we swerved four times into approaching traffic, each time more out of control than the one before, and it was only through grace and providence that each time we spun across the middle paint line there was no other car in position to slam into us.

After the fourth wild, wide skid across the lanes Pongi said "Here we go," and we crashed through a guardrail and over a steep embankment. In the diary entry for that night, I wrote that I had been calm, but sad. I knew these were likely the last seconds I would be alive. I had not told my parents I was going out of town for the day.

The car bounced three times and finally came to rest in a ditch. We looked at each other; somehow we were all right. We climbed out of the car and into the blizzard.

Someone in a nearby house must have dialed the telephone operator; within ten minutes two ambulances, a police cruiser and a tow truck arrived. We stood in the snow; a news car from Channel 10 television in Columbus pulled up—the station evidently heard about the accident on the police radio scanner, and it must have sounded bad enough for them to have dispatched the cameraman.

It's just a story for us now; just something we talk about from time to time. Because it turned out the way it did, we can do that. But one car barreling at full speed the other way through the snow at the wrong moment, one two- or three-second delay in Pongi grappling the steering wheel to the right or to the left to bring us back across the road, one different-vectored bounce when we went through the guardrail, a bounce that would have flipped the car onto its roof . . .

The combination of factors that, thirty years later, ended up bringing Jan and Dean, Jack and me together at a Twin Cities concert was one way—one sunny way—the fates can conspire to change lives. But there are other ways. Any shift in the factors of fate that night on the highway . . .

Jack and I, outside his house, stood with the boxes in our arms next to his car, and as bad as things were for him, as terrible as all of this was, there had been a chance that we never would have made it to here. That everything in our lives—everything good, everything bad—that had occurred between that night and this morning would simply never have taken place.

I don't know if the others prayed when they got home that night. I did.

ALTHOUGH EVEN ON A NIGHT LIKE THAT—when you are young—there are reminders of unsuppressed joy.

The Channel 10 cameraman? The one who showed up in the snowstorm?

As soon as we saw him we knew that although we were alive, we now faced a different problem. If our parents turned on the eleven o'clock news and saw us next to Pongi's crashed car . . .

Dan sprang into action. I have no idea how he knew to do it.

When the cameraman started filming, Dan stepped right in front of his lens—and held up his middle finger.

The cameraman tried to shoot pictures of the car, and of us, tried to shoot pictures of the police officers checking out the scene.

But Dan was dogged. Everywhere the cameraman moved, Dan moved. Every direction the lens pointed, in front of it was Dan's finger. I don't think there was a single exposed frame of film that night that didn't have, as its main element, Dan giving the lens the finger.

He knew that Channel 10 would never put on the air a piece of film showing a kid giving the finger. It was a sagacious decision, and one Dan made on the spur of the moment; in that era, there was no way the film was going to be broadcast. (Today, a cameraman would not only find a place to air such film footage, he'd probably pitch it as the pilot for a reality series. Dan would be a star on MTV.)

That night, for us, there would be no TV appearance; Dan had saved us from our parents' wrath. The cameraman finally muttered something and got back into his car and drove away; the tow truck pulled Pongi's car from the ditch and he managed to drive at a crawl back to Bexley.

In one night: the most fearful of possibilities, and laughter that was a mixture both of relief and of the dawning recognition that in Dan there was some nontextbook variety of raw genius. Pongi had to tell his parents, but ours never found out. Dan had made sure of that. We asked him how he'd thought of it. He'd said: "You guys were just standing there. *Someone* had to do something."

NOT THAT SUCH AN INSTINCT NEVER REappeared.

One November when all of us were in our early fifties, and my mother was in the hospital in Columbus and I was spending time there, I was supposed to do a live television commentary late at night for a national cable newscast. Hospital visiting hours had ended, and I had no convenient way to get to the local television station that had agreed to provide the remote facilities.

So Jack and Pongi said they'd take me up there. The station, it turned out, was Channel 10—the same station whose camera crew, more than thirty-five years earlier, had showed up at the guardrail scene.

We arrived at the studios—I was doing the remote from one of the Channel 10 news sets, the national cable show was paying them to provide the cameras and the set and the technical personnel—and Jack and Pongi and I walked back to the area that had been set up. I put the earpiece in and fastened the microphone to my tie and sat in the chair facing the camera.

Off in another part of the room, some Ohio State football coaches were taping a preview show for that Saturday's game. Jack and Pongi recognized the coaches, and the sportscaster who was interviewing them, and watched intently. I could tell that this was sort of exotic territory for them.

Then the time arrived for my commentary to go up on the satellite—it was one of those setups where the anchors in the distant city asked me questions and I provided unscripted news analysis. I was hearing the anchors but could not see them; there was no monitor beneath the camera.

What I did see out of the corner of my eye was Jack and Pongi, staring silently at me. I didn't break eye contact with the lens, but even as I was talking I was thinking: What a stupid skill to have. Speaking to a glass lens as if it were a person. A lot of the time, I was more at ease doing that than talking to people face-to-face. It was just something I had developed over the years: a preference for this kind of communication, through a faceless lens or through a keyboard beneath my fingers—a preference for that over the kind of communication human beings are supposed to be best at.

There were exceptions, however, and Jack and Pongi were among them. The people with whom I never felt even a twinge of unease were the same people with whom I'd first spent so many days and nights, and even though my words on this night were about some news item that undoubtedly was being presented as overweeningly consequential, the news item would be forgotten within a week. I knew it even as I talked.

And to the left of the camera, in folding chairs, were Jack and Pongi; I saw them for the entirety of the commentary, although I don't think they knew it, and all I wanted to do, for their benefit and amusement, was what Dan had done to the Channel 10 camera on the highway all those years before.

I knew they'd love it; I knew they would fall off the chairs. If I'd flashed my middle finger at the camera they would have gotten the significance right away, it would have been far more real a moment than whatever alleged insights I was providing about the story in the news. It was another of those *How did we get here?* instants—it seemed that we were at the guardrail in the snow one minute, and then suddenly we were here, and what happened to all the years in between?—and as I spoke to the flat and impassive lens, spoke with more emotion and animation than I would to a person, I wanted to do to the lens what Dan had done to the lens. But of course that would have been suicide; the people on the other end would not have understood, nor should they have, and of course I didn't do it. Making Jack and Pongi laugh would not have been worth it.

Maybe.

EVEN WITH JACK, THOUGH, SOME THINGS were difficult for me to say.

We were at his house the same day we'd unloaded his car trunk. There was an underlay of unreality that was building in these months; we each knew how the story was going to end, there wasn't much doubt about it by this point, yet to dwell on it was something neither of us wanted. So much was going unspoken, by our own choice.

We went out back, to the alley behind his garage, and a garbage can—apparently it had been left in the alley early that morning for city sanitation crews to dump—was there. Although the contents had by now been poured out, it was still dirty inside—it was a garbage can, after all.

Jack walked over to take a look at it and then called, "Greene, do me a favor."

I joined him and he said, "Carry this garbage can back to my garage."

I'm not a guy—how should I say this?—who has a history of being overly devoted to the idea of doing chores. It had been a running joke with Jack and me for years. I'm not the most helpful person around the house—any house.

Those friends of Jack's who had worked in his basement to put up the shelves? Think of the opposite of those guys. Jack, in his whole life, would never have asked me to take a garbage can anywhere. It wouldn't have occurred to him.

But he was asking me now. And we both knew why.

The chemotherapy had lowered his resistance dramatically. He was susceptible to all kinds of infections. The last thing he needed was to put his hands anywhere near a garbage can.

He knew it and I knew it.

And neither of us said a word about it. At least not a serious word.

I carried the garbage can to the garage and then shot a look in Jack's direction. It was a look you would give to someone who had just requested that you climb Mount Everest twice in a row.

He broke into a grin.

"I know you didn't want to do it, but it may have saved my life," he said.

"That's all very nice, but it ruined my day," I said.

We stood in his driveway, looking at each other, holding on to the moment.

HE'D HEARD SOMETHING.

It was while I had been back in Chicago, he said; someone told him about it.

"They finished fixing the track," he said. "As soon as I heard, I went over and looked."

And . . .

"My heart skipped," he said.

Not the way your heart skips when it is filled with new romance; not the way your heart skips when you're startled.

"Remember those dirty black cinders the track was made of when we were kids?" he said. "If you fell down you'd get those things stuck in your knees. I don't even want to think about where they got the cinders."

We'd walked the track many times since I'd been coming to see him these months, but apparently there still had been some final work to do, and now it was ready.

For the future.

That, he told me, is why his heart had skipped.

He wasn't going to be around for it.

Everything had sunk in when he realized that. The city had gotten the track ready for the future—like all renovation projects, that was the goal. The new track was meant to be useful and pleasing for new generations of the people in the town, starting right now.

And he wasn't going to be one of them.

HE DIDN'T FEEL WELL ENOUGH TO GO FOR a long walk and his stamina was low, but he asked me if I would go look at the track with him this one time.

So we walked there, and we stood on the track, and he talked some more about the filthy cinders on which we once ran.

"They probably were a health hazard, but you have to

keep in mind, that was the era of blood brothers," he said.

"What do you mean?" I asked.

"Don't you remember—little kids would prick their fingers, and hold their fingertips together at the cut parts, so their blood would mix?" he said. "That was how they showed their friendship. By becoming blood brothers."

"That's one you don't hear about these days," I said.

"If a kid came home now and told his parents he'd become blood brothers, the parents would probably call the National Institutes of Health," Jack said.

"If not the National Guard," I said.

He took a long look around. "Let's go see if that water fountain is still there," he said.

It had been this crummy ceramic thing—hard and white and discolored by rust stains, bolted to a brick wall near the side of the football grandstand. The stream of water that came from its scratched metal spigot was always weak but surprisingly cold. You had to really work to turn the handle; you thought if you twisted hard enough you could make the water arc a little higher, so that you wouldn't have to lean so low to drink. But the parabola of water perpetually stayed the same, too close for comfort to the ceramic base.

"It felt like spring," I said. "I don't know why. I think of spring when I think of that fountain."

"I know why," Jack said. "We'd always go drink from it when we were little and we would be watching the high school baseball games after school. The varsity diamond used to be there—the fountain would be behind the right fielder."

At those baseball games we wouldn't wear jackets, because it was March or April and the sun was brilliant in the sky. We were seven or eight, and we'd watch the high school varsity, and we'd constantly be running out to that cruddy little fountain. When the sun would start to go down, we'd get chilly—the earlier warmth had been a tease, and we would shiver in the late innings.

But we'd never learn our lesson; next game, we'd be back without our jackets again, running together to the fountain again, sure that one spring day the sun would set and we'd still be warm enough. And one spring day we would be right. The warmth would last.

We walked toward where the fountain had been, and without saying a word to me Jack did something.

He dropped to one knee next to the football field. And he touched the grass.

He let his hand rest on top of it for two or three seconds. Then, without explaining, he stood.

I knew we'd never be coming back.

BEFORE GOING HOME WE STOPPED AT THE ABCDJ brick, and at Audie Murphy Hill.

This was the trip to the hill when he said it was too steep to climb. When—as when he was a boy—the gentle little rise of grass looked, through his eyes, to be too much.

Twenty-one

If JACK'S DAD EVER OWNED A HOME-movie camera, I didn't know about it.

My father was obsessive about shooting home movies; he had one of those Bell + Howell handheld silent film cameras, and he took movies at every family occasion, on every vacation trip, at every holiday gathering. He stored the movies in metal film cans, with pieces of white tape as labels on each one, the subject and date of every movie neatly cataloged in his own handwriting on the strips of tape.

I didn't much like the movie-taking routine in our family. The filming seemed to supersede the experiences; posing for the home movies felt, to me, like a stagy substitute for the experiences themselves. The shooting of the movies shoved the real occurrences aside.

I never saw a movie camera at Jack's house, I never saw

a screen or a projector. If in fact Irvin Roth didn't take home movies, that would make sense; as a first-generation American, he may have been a little slow in jumping into suburban postwar customs.

It does raise the question, though: Where do all the memories go? If you don't make an effort to save them, do they just drift away?

AT THANKSGIVING VACATION OF MY FIRST year off at college, my parents did something in honor of my friendships.

On the Friday night of the vacation—the day after Thanksgiving itself—they held a dinner party at a restaurant. They had sent out invitations to Jack, Allen, Chuck and Dan.

We gathered at the restaurant, and from the first moment everything felt stiff, different. Part of the reason was the presence of my mother and father at the table; they were the hosts, yes, but it's a little awkward for five guys in their first year in college, their first year away from home, to sit in coats and ties and talk candidly about their new lives in front of one of the guy's parents.

Even taking that out of the mix, there was something in the air that night that knocked all of us almost imperceptibly off our gaits. We all had begun our separate journeys away from everything that had been, on our way to everything that would be. I was off in Illinois, at Northwestern; Jack was down at Ohio University; Allen was at Miami of Ohio; Dan was at the University of Arizona; Chuck was in Columbus at Ohio State. Our frames of reference, as never before, were distinct. Our five lives were suddenly filled with names and places that didn't overlap.

I recall the dinner as being vaguely melancholy, with long silences. It made me think about the college sports teams whose stars Jack and I had interviewed for the junior high school *Beacon*. The first teams you fall in love with you think are going to last forever. You learn soon enough that it's not so, but when you are young and those first teams do break up, you can't quite figure out the feeling. Jack and I had loved the Lucas-Havlicek-Siegfried-Nowell-Roberts Ohio State NCAA championship basketball team; when the seniors all graduated, when the new Ohio State team, with a player named Gary Bradds at center, took to the court at St. John Arena, it was disorienting to us. Especially when the crowd cheered just as loudly for that team as they had cheered for the Lucas team; especially when the Bradds team began to win consistently.

The first things that matter to you will not remain as they were; they can't. It's fundamental, and—once you learn it that initial time—you realize that it must be so. You look around the basketball arena and ask yourself if anyone else is thinking: Where's Lucas? If, even for a second, as the people in the crowd are on their feet and roaring for the newly constituted team, they are thinking the same thing you are: What happened to what was here before?

Probably not; you only get that sensation once, and when the lesson has been learned, inexorable changes never again seem as remarkable to you. At Thanksgiving of the first autumn that the five of us had been waking up in different cities every morning, we got through the dinner party, and I don't know if the other four were feeling as wistful about it as I was. We weren't the crowd, the people in the stands: We were the teammates. If the receding away of the first great team is jolting for the spectators, what must it be like for those who play the games? How must Lucas and Havlicek have been feeling that first year out? If the people who had watched them with such passion and seeming devotion just a year before had moved on to new loyalties, what about them?

It was very nice of my parents, to do that: to send out those invitations, to try to preserve for me that which had already begun to fade.

DURING ONE OF THE JUNCTURES BETWEEN trips to see Jack—one of the stretches when I was back in Chicago—I was going through piles of old mail that had been accumulating: magazines, bills, brochures, opened letters.

I was throwing almost everything away. I'd glance at each item, then discard it. My goal was to clean it all out.

I picked up one empty envelope, something that had arrived and been opened many weeks before, and as I was about to toss it into the trash I glimpsed at it and I stopped.

The envelope had been addressed to me by Jack.

He'd sent me a newspaper clipping or something—he was always doing that, he always had, that hadn't changed once he'd gotten sick. Whatever had been in the envelope, I had looked at the day it arrived. But the envelope itself was still lying around.

There was that handwriting of his—blocky printing, angled slightly to the left, handwriting I had been seeing for more than half a century. Handwriting I had seen when it was brand-new. We learned to write our names together, in the same classroom.

The year that we'd gone away to college—the year of my parents' Thanksgiving vacation dinner for all of us—Jack and I had written back and forth to our separate cities, to our different dormitories. Not a lot; you're busy that first year away, you're making new friends, you don't spend much time reaching for the old.

But whenever a letter from Jack would arrive in my dorm mailbox that year, the envelope felt like home before I even opened it, because of that handwriting. We had never before had occasion to write letters to each other; suddenly we were doing it. Most likely the feeling is different for first-year college students now; the e-mails and instant messages fly back and forth, the cell-phone nationwide-minutes packages and text-messaging options may make it seem as if no one has really departed from anywhere. One arriving e-mail address in a computer screen in-box appears like any other, as far as font goes.

Sitting in Chicago, with the empty envelope in my hand, I contemplated Jack's writing—my name and address, in his hand—and I knew it was a sight I might never see again.

I thought of him kneeling to touch the grass of the football field.

I kept the envelope, of course. I put it somewhere safe, for whenever I might need to take a look.

"SO THEY GAVE ME THE OXYGEN THING and . . . Hold on, Greene."

In Chicago, I sat with the phone to my ear.

I was getting ready to go back and see him, and he was explaining to me that the oxygen tanks and the tubing had been delivered to his house—"just for when you think you might want to try it," his doctors had said, but he got the message very clearly—and then he'd told me to wait. A minute or so passed.

"Sorry to do that to you," he said when he returned to the line. "That was Chuck."

Chuck was on a trip out of Columbus; he'd phoned, and the call-waiting had clicked in, and Jack had spoken with him before coming back to me.

Nothing extraordinary about that. It happens every second of the day on telephones in every city in the country.

Except, that question: Where do the best things go? What do the teammates think about, once the first, dearest team has had to disband?

All these years later, and at the same moment, we're still calling Jack, just as we always did. I don't know how you put a price on that. All these years later, and it's still us.

"I hate to have made you hold," Jack said. "What was I saying?"

"The oxygen," I said.

Nothing extraordinary about any of this. Except everything.

ONCE WHEN I WAS IN MY FORTIES I WAS on the road on business, and I had finished up for the day, and I had my evening planned.

I was looking forward to it. There was a basketball game that I wanted to watch on television; the place where I was staying had a little sandwich shop, and I'd asked them to make me a submarine sandwich with ingredients I had picked out myself. I'd also bought a six-pack of beer and some barbecue-flavored potato chips, and I'd put the sandwich and the beer in the room's refrigerator.

The rooms in the hotel featured both refrigerators and microwave ovens; my plan was to heat up the sub sandwich one half-sandwich at a time, sit back and watch the game, eat the hot sandwich and the potato chips and drink a few of the bottles of beer. To just relax.

Which I began to do—good sandwich, good game, cold beer.

And the phone in the room rang.

It was Jack—this was years before he got sick, this was just a call to check in and say hello.

I told him I couldn't talk. I didn't say why. But I wanted to eat the sandwich before it got cold, and drink the beer before it got warm, and watch the game I'd been anticipating. I told him I'd call him the next day.

As soon as I hung up, I'd thought: Now, why did you do that? It wasn't anything earthshaking that he was calling about, but what was so important about your sandwich and your ballgame that you couldn't talk on the phone instead?

And I remember thinking to myself, even then:

There may come a time when you want to talk to him, and you can't.

ONE NIGHT IN CHICAGO, AS I WAS GETting ready for another trip to Columbus, I called his house and no one answered.

This was not unprecedented; if he or Janice were talking to someone else, or if they had visitors, there were times when they didn't pick up. So I left a message, waited an hour, and called back. Same thing. I knew he wasn't going to be going out to dinner that night—he'd had chemotherapy scheduled for earlier in the day, and he stayed home at night after chemo days.

So I was concerned, especially when he didn't respond to either of the voice mails I left.

The next afternoon I called and he answered, sounding low.

"I tried to reach you last night," I said.

"I know," he said. "I saw the call coming in on the caller ID. But I just didn't feel like talking."

That was Jack: He wouldn't even tell a small lie. Most people would say they hadn't known you'd called, or were sleeping . . . Jack only knew how to tell the truth.

But this was a first, as far as I knew: choosing not to pick up the phone when I called.

"I was just so down last night, I asked Janice to let the phone ring and for us not to talk to anyone," he said.

"What happened?" I said.

"I went to have chemo yesterday morning, and they did some blood work, and I have some numbers that aren't good," he said. "They have to be higher for me to be able to start chemo again."

"What does it mean?" I said.

"Nothing good," he said.

"Does it mean they can't try anymore?" I said.

"There's this shot they can give me," he said. "Sort of a booster shot for my blood. If it works it can get the numbers high enough for me to have chemo again."

"Boy," I said. "Did you ever think we'd be talking about getting a booster shot for something like *this*?"

"I know," he said. "And if it works, the good news—the

good news—is that I get to have chemotherapy. That's the payoff."

"Some jackpot," I said.

"You know," he said, "it's getting so that I don't even like it when people say 'What's going on?'. I find myself saying, 'Oh, my pulse is down,' or, 'Well, I need a booster shot.'"

It was the second time in recent months that he had said something like that to me: that he was made ill at ease by people asking him how he was doing.

"They're just hoping that you'll say everything is going great," I said.

"I know," he said. "And I feel like I'm letting them down when I tell them it isn't."

I SUPPOSE I WAS ONE OF THOSE PEOPLE. Holding out hope that one day he'd pick up the phone and say the clouds had all lifted.

There used to be a beer-and-wine carryout place called the T.A.T., at the corner of Broad Street and James Road in Columbus. Before we were old enough to buy alcohol for ourselves, we would drive to a parking lot next to the T.A.T. and look for old guys who appeared as if they might be a little down on their luck. Such guys weren't in short supply on that block, and many of them were patrons of the T.A.T.

We'd offer to buy them some beer if they'd buy us some, too, and bring it out to us. Some would, some wouldn't. Jack and I would stand there, with Chuck or Dan or whoever else had volunteered to make the beer run, and we would wait for the right-looking kind of guy to approach the T.A.T. When one did we would hold our breaths and walk toward him and hope he wouldn't turn us down, or go inside and tell the clerk about us, or turn out to be some kind of cop.

That was hope for us, on those nights—dumb as it sounds, that's what qualified as hope: that we'd find some guy to buy us beer. It's hard to conceive that our lives were ever as uncomplicated as that, that at least for a few minutes on a slow-moving evening we would genuinely and enthusiastically get our hopes up that our scheme would work.

Now hope consisted of that microsecond on the telephone when I would call Jack and wait to hear if just maybe his voice saying hello would be buoyant and full of cheer.

"So anyway," he said. "That's why I didn't answer the phone last night. Believe me—you didn't want to talk to me."

"I WOKE UP IN THE MIDDLE OF THE NIGHT and I was having trouble breathing," he said.

This was a day or two later. He had started to use the oxygen.

"Did you sit up all night, or were you able to get back to sleep?" I asked.

"I woke Janice up," he said. "I told her that I'd been thinking about things I shouldn't be thinking about."

He said to me that all through our lives, we had taken it on faith that we were moving across a long continuum. We knew when we were at the beginning of our lives, and we knew when we were past the beginning, and we thought we knew when we'd reached the middle. That was basic—the path we were on.

"But what happens when the time comes when you aren't anywhere on that path?" he said to me. "What happens when everyone else is still on the path, and you aren't?"

"Well, I guess that's what happens to everyone, eventually," I said, lame as could be. He had caught me by surprise.

"But after you're not on the path any longer," he said, "does it seem like you didn't even exist? Not that you were never alive—but all of a sudden do people think of you like a person who lived in 1913 or something?"

It wasn't the dying he was talking about—it was the not being on the path. Not being in the game. Everyone he knew and loved would still be there. And he wouldn't.

I wanted to tell him that this—the pain, the fear, the sadness—wasn't the real part. I wanted to say that he wasn't going anywhere—that the life he had lived was the real part, that he wasn't leaving, that he was always going to be here. That what he—what all of us—always had together was what would endure.

I had no way to say that without sounding like a person he didn't know. Even though I believed every word of it.

What I said instead was that I would fly in the next day.

AFTER MIDNIGHT ONE NIGHT DURING Jack's struggle I woke up and I went to the kitchen for some water.

I sensed something was amiss. I looked and my son's door was open; he wasn't in his bed.

He'd been there earlier; it was late in the summer after

his senior year in high school, just before he was to leave for college, and he'd already come in for the night and we had spoken.

He and I were living alone in the house; I hadn't thought about it before, but I suppose, in the time after my wife's death, I was playing the Irvin Roth role.

When Jack and I were his age, if my parents had awakened in the middle of the night to find I was not home, there was not much they could do about it until they heard from me.

But there were no cell phones then.

I dialed my son's cell-phone number; he picked it up on the first ring.

"I didn't want to wake you," he said. "I'm outside having a cheeseburger."

"What do you mean?" I said.

"Wee and I were talking on the computer, and he sent me an IM saying he was hungry, and did I want something from Wendy's," my son said. Wee was short for William—one of his two best friends. I don't know where William's family came up with the nickname.

"And you went to Wendy's?" I said.

"No, he brought the cheeseburgers over here," he said. "It's a beautiful night out. We're just sitting on the curb and eating them. I'll come back in soon."

It was, indeed, a gorgeous night outside—just the kind of night for a cheeseburger with a buddy you like the most. The kind of night to sit outside on the edge of the curb, knowing without really knowing it that all of this will be gone before long, that your best friends and you will scatter to different places soon enough, and that nights like this will start to be much more rare. Take them when you can. They're perfect.

Like Jack's father, I had never been one to shoot home movies. The best pictures, the most lasting and most vivid, are the ones you see with your heart. They never fade, and they never get old.

The best things do not disappear, ever. That's what I wanted to say to Jack. I knew I wasn't wrong.

Twenty-two

When I arrived in Columbus he was in the hospital.

He'd had another bad night trying to breathe. The doctors wanted to take a good look.

He'd be sleeping there that night; with any luck he would get to return home in the morning.

I was too late for visiting hours.

Chuck, Joyce and Janice called me at my hotel; they had left the hospital, they hadn't had a chance to eat dinner, and although it was late they said they were going to get a sandwich at a place called the Old Bag of Nails Pub, out on Nelson Road near Alum Creek. I said I thought I could talk the hotel's van driver into giving me a lift.

When I got to the restaurant it was mostly empty and the three of them were in a booth. We ordered drinks, and we'd been talking for about ten minutes when I saw a smile sneak onto Janice's face.

At least I'd thought it was Janice.

They'd pulled the trick on me.

Janice had sat next to Chuck, pretending to be Joyce; Joyce had sat where I was assuming Janice would be. They'd wanted to see if I could figure out that the twins had been switched.

I shook my head in exasperation; they'd managed to pull it off.

But I was an easy one to fool, on this night. My mind was somewhere else, as were theirs. They'd reached for a little laughter; it had worked, for a second. None of us knew what else to do.

IT WAS THE HALLOWEEN PART OF AUtumn—you could almost feel winter gathering somewhere off in the distance, forming in the atmosphere high above Manitoba, taking its time before it would embark upon its annual journey to this part of Ohio, where it would linger until March or April.

Now the nights were a shifting mixture of warmth and chill, like the temperature inside a stall shower ruled by a balky knob imprecisely adjusted. The children of the town would within days be out in their trick-or-treat costumes, and the newspapers were already carrying stories about neighborhood safety patrols, and police stations where parents could bring suspect items, and plans for a daylight-only Halloween in some suburbs. The world itself had turned considerably darker in the years since our own Halloweens, and the last thing needed, it seemed, was artificially induced fear. Real fear was not in short supply.

The first Halloween that Jack and I were allowed to go door-to-door on our own, my mother stood at the end of each block and watched us from afar, or what she hoped, for the sake of our wished-for independence, we would believe was afar. I was dressed as a skeleton, in black tights and a black sweatshirt with the bones painted on with greenish-white paint that glowed in the dark (it glowed like the bones in your feet used to glow in those shoe-store fluoroscope fitting machines; I wouldn't be surprised if the bone-paint on my trick-or-treat costume was made of some of the same ingredients used to manufacture the atomic bomb). Jack was dressed as a ghost, draped in a bedsheet into which his mother had cut a hole for his head.

So with my mom waiting on the corners, watching as we passed beneath the streetlights, Jack and I would ring one doorbell after another, politely informing each person who answered that we had arrived to scare them. All over town it was a slow-moving parade up and down the streets, Stanbery and Sherwood and Montrose and Ashbourne and Dale and all the others: fairy princesses and football players, cowboys and gangsters, all the dressed-up children, all the moms waiting no more than fifty feet away.

My mom had bought me a goblin-themed paper bag over at Rogers' Drugstore, and that is what I carried to collect my candy; Jack's mom had given him an old Lazarus shopping bag, one of the big ones with the sturdy rolled-paper handles. Door by door we made our way up and down the blocks, and the most fearsome thing we saw came from inside a house, not out: A man opened his door with fury and, I now understand, despair in his eyes, he was drunk as could be, he was the father of a boy we knew, and in his home that evening, as, I suspect, on many evenings, he was intoxicated, reeking of alcohol. He didn't say anything to us, or not anything we could decipher—just stared at us through those hurting eyes and murmured angrily, then handed us each some candy. When we saw his son in school the next day we said nothing.

That was the exception, though; at the other houses we were the ones who were supposed to be scary, Jack in his ghost clothes, me in my skeleton. It was a night that felt very much like this night in autumn, as Chuck and Joyce and Janice and I sat inside the Old Bag of Nails, with Jack a few miles away in his hospital bed, trying to breathe. A night very much like this one, designated to frighten.

MAREN WAS HOME NOW, FOR THE DURAtion.

That's not the word she used to describe it, but it was the accurate word. She had asked her employers in New York if she could take a leave of absence to be with her father. They had said yes. She was still doing some work, via computer and long-distance telephone, while she was in Ohio, but she was going to be here as long as Jack was alive.

He was always very proud of her, and also always a little private about her. Jack was a guy who shared virtually everything about his life with everyone. About Maren, he was more close to the vest. It was as if the depth of his feelings for his daughter—his overwhelming love for her—was something almost religious in its importance to him, something the specifics of which he wanted to save for himself and Janice.

Maren was a bright and pretty young woman who, in the middle of her twenties, was fitting right in with the exhilarating new life she and her contemporaries were living in Manhattan. But the decision to come home to be with Jack had really been no decision at all: This was where she belonged, and she'd hardly had to think about it.

"Every day when I was little and I would go to school, I would open my lunch bag and there would be a note from him," she told me. "Just a little message he would write on a piece of paper every morning, so when I was eating lunch I would see it. He told me he loved me every day."

She was sleeping in the bedroom of the house he had built for Janice and her and him, her bedroom just a few steps away from the room where he and Janice slept. When he needed help in the night, she was within the sound of his voice.

SHE STILL HAD BEEN LIVING IN NEW YORK during his visits to Sloan-Kettering.

Jack's mood during those visits had surprised me by its lack of gloom, especially in light of the fact that the news he was hearing was not hopeful. But because Janice was there, because Maren was there, he was able, on some level, to convince himself that this was an unexpected little vacation in Manhattan, and that he should make the most of it.

He'd called me during one of those New York visits—he was staying in an apartment that a man with whom he'd done business had let him use. The business associate was going to be traveling during Jack's trip to Sloan-Kettering, and he had left his keys with the doorman for Jack and Janice.

"You should see this place," Jack had said to me. "Can you hear the sounds of the city? I'm sitting out on this wonderful balcony . . . the apartment is decorated with souvenirs my friend has brought back from business trips all over the world. The most beautiful things . . ."

And meanwhile he was trying to push out of one corner of his mind, for a few minutes or a few hours, the reason he was in New York in the first place.

During his most recent appointment at Sloan-Kettering, Jack had told me, the doctor once more said that the treatments he was getting in Columbus were the appropriate treatments—that there was nothing the Sloan-Kettering experts could suggest that would give Jack a significantly better outlook.

"I thanked him," Jack had told me, "and I asked him if I would see him in six months.

"He told me, 'Maybe. Maybe not,'" Jack said. "He said he hoped so."

When Jack returned to Columbus, he told me, one of the doctors said to him: "You know, this is not a curable disease."

It was too advanced by now. They were doing their best to make him understand that.

"FOUR TICKETS, PLEASE."

I could barely make out the words.

He was in bed when I got to his house—this was the

day after the Old Bag of Nails dinner, he was home from the hospital, the doctors had been able to get his breathing stabilized. He was in a hospital-type bed that had been brought to his bedroom. It enabled him to sit up more easily.

The oxygen tubes were in his nose. But he was smiling as he said those words to me: "Four tickets, please."

People had been bringing him DVDs. This had been going on for months; people thought he might enjoy watching movies, so they would buy or rent DVDs and drop them off for him. He appreciated the thoughtfulness behind the gestures, but it was getting to be too much, all the DVDs; his desire to be entertained was limited at best. "A lot of *Pirates of the Caribbean*," he had said to me.

Now there were more DVDs piling up, and what he had said to me through the oxygen tubes about the four tickets was a signal between the two of us—a signal about the kind of entertainment we had once sought, knowing we would probably fail.

There used to be an "art theater" in Columbus called the Parsons—it showed dirty movies in the era when dirty movies consisted of little more than films of women sitting around in white bras, smoking cigarettes and drinking martinis. Still, that was more than we had ever seen before, and we hoped to secure admission to the Parsons.

There was an age requirement—it may have been eighteen, it may have been twenty-one, whatever the age was, we hadn't reached it yet. But we really wanted to see what it was they were showing on the screen in there.

There was no way that was going to happen before one of us got a driver's license; the Parsons was in a part of town that was iffy at best, even in the daytime, and the Parsons didn't show its movies in the daytime. This was not a neighborhood to which we were going to ride the bus.

Pongi was the first of us to turn sixteen, the first to get his license. On a weekend night that winter, with Jack and Dan and me in the car, Pongi drove to the block upon which the Parsons sat, parked around the corner, and, with Dan and me, waited while Jack went on his mission.

We had worked out a plan. Jack, at fifteen, looked a little older than the rest of us, and his voice was a little deeper. He was going to round the corner to the Parsons, walk up to the box office in as seemingly casual a manner as he could muster, and attempt to buy tickets for all of us. How we were going to use those tickets to get in was a question that would have to wait until after the purchase was made.

So he did it. He strode purposefully to the box office, dropped his voice as many octaves as he could, and said to the man:

"Four tickets, please."

We heard all about it a few seconds later when, defeated, he returned to Pongi's car and said, "Let's get out of here in case the guy calls the cops."

So as Pongi squealed away, Jack played back for us exactly what had happened. How he had said it with as much authority as he could—"Four tickets, please"; how the man had barely looked up from his magazine to see this fifteen-year-old kid standing in front of him; how the man had asked why the boy needed four tickets, and where the other three would-be patrons of the Parsons Art Theater were; how the man had shooed him away. Whether the man had literally laughed out loud at him,

Jack was not sure, but the laughter was at the very least implicit.

It had been worth a try. That's what we had told ourselves. Who knew what waited inside the doors of the Parsons? Bras on-screen, and maybe even more, or less.

Now his friends were bringing him *Pirates of the Caribbean* for his movie-viewing pleasure. They meant well, even though he was feeling a little condescended to.

"Four tickets, please," he said through the oxygen tubes, and we both shook our heads.

DESPITE HIS CONTINUING FRUSTRATION about what was happening to him, he was grateful for all the people who were doing everything they could think of to make him understand how much they cared.

"I got the best card in the world today," he said, reaching toward his night table so he could hand it to me.

"From Miss Barbara," I said.

His hand paused in mid-air.

"Who told you?" he said.

"No one told me," I said.

"Then how did you know?" he said.

"You said you got the best card in the world," I said. "Who else is going to send you the best card in the world?"

He handed me the card. From Miss Barbara.

Our kindergarten teacher was well into her seventies now. She never forgets anyone. And she always sends the best cards. It wasn't that daring of a guess on my part.

"Oh, Greene," he said. "I just hate this."

Twenty-three

He was sitting in front of a computer the next time I saw him. This was the following morning; he was wearing sweatpants and a sweatshirt, sort of de facto pajamas—the computer had been set up in a guest bedroom, so he could walk back and forth between it and his bed.

The oxygen was with him, as it would be most of the time from now on.

As I entered the room he looked up, like an executive greeting someone in his private office. If, for some reason, the executive had heavy cylindrical green metal tanks stored against one of the walls.

"I just did something nice for my family," he said.

HERE IS WHAT HE HAD DONE:

He had written letters to the people with whom he did business. In the letters, he proposed that for deals he had set up in recent years—deals that continued to provide revenue to him—his associates agree to something.

Jack wanted them to promise that the money that should go to him would go to Janice instead after he died.

His theory was that there was already an agreement that he was to receive a certain percentage of the revenues from each deal—royalties, as it were. The way he explained it to me, that was the basis of how he made his living. He set up certain business arrangements involving products and distribution. If the arrangements worked out, he was paid from the proceeds, until there were no proceeds. When deals ran their course, he would begin the process of setting up new deals.

But now, he knew, there would be no new deals.

"I'm just trying to get everyone to assure me that after I'm gone, they'll take the money they would have sent to me, and send it to Janice instead," he told me.

It seemed only fair, he said.

That's what he was doing sitting in front of the computer, with the oxygen being pumped into him:

"Something nice for my family."

I HAD NO IDEA WHETHER IT WAS GOING to work.

I had my doubts.

Jack's first instinct, all his life, was to trust everyone.

This drove some of his friends—Chuck, in particular—nuts.

Unconditional love, in romance novels, is an impelling quality. But unconditional trust, in the world of business, can lead to bankruptcy, among other dire destinations. Chuck could never have been accused of viewing the business world through a gauzy, my-fellow-man-and-I-walk-hand-in-hand-in-fields-of-daisies lens. Jack, on the other hand, never stopped believing—truly believing—in do unto others.

This just might put it to the test.

He was not a wealthy man.

"I think I'm pretty close to getting everyone to agree," he told me. "I'm just waiting to hear back from some people about a few things."

I knew that there were lawyers and accountants who could blow Jack's theory apart in an instant; I knew that once a man was dead there were countless ways to quite legally close out his accounts, and start from scratch. And he was in no position, right now, to drive a hard bargain.

From downstairs, I could hear Janice and Maren talking in the kitchen. Since Maren had come home to be with Jack, the house had the sound and feel of family again. There were always voices.

His wife and daughter weren't aware of what he was doing up here.

But he was doing it for them.

"I CAN BE IN TOUCH WITH EVERYONE I need to from here," he said.

The world was, in fact, coming directly into this little room on Bexley Park Road. Through his computer, he could correspond with just about anyone anywhere, and know that the letter would be delivered in an instant. He had his cell phone on the desk next to the landline phone; he was reachable both ways. The television set, with its multiple cable news channels, was ready to show him events anywhere on the planet at the moment they happened. And the amazing thing was not that such devices existed—the amazing thing was that they had come to be considered standard-issue. They weren't the province of kings. They were available to anyone, as common as tap water.

"Remember when Sally Flowers used to send her best wishes to 'all our shut-in friends'?" I said. Sally Flowers was an early-years-of-television local daytime host on Channel 4 in Columbus—a pleasingly hefty, dark-haired central Ohio woman who favored housedresses on the air, and who strongly resembled Elvis Presley's beloved mother, Gladys.

"Sponsored by Moore's Stores," Jack said.

"She'd say it every day," I said. "'And to all our shut-in friends out there. . . .' It means something different now."

"Why?" Jack said. "People are still shut in."

"Maybe physically," I said. "But you have access to

more right here in this room than you do when you walk out your front door and onto the street."

All the finest stores and shops anywhere in the country, all the books and musical recordings and movies anyone could ever desire, the cataloged knowledge of the ages, advice from experts in every field . . . all at anyone's beck and call: all at Jack's beck and call, from the same keyboard he had been using to try to do something good for Janice and Maren. The news of the minute, in sharper and closer focus on the television screen than if Jack had been able to get onto an airplane and try to witness those events firsthand. Friends and associates all across the nation, available to speak within seconds in their homes, their offices, their cars, while walking down the street. This transformation—the world delivered to one's room—had happened so gradually, yet paradoxically so quickly, that its impact was still hard to measure.

"Talk about 'Four tickets, please,'" Jack said. "Have you seen some of the stuff that's out there on the Internet?"

"I know," I said. "I don't think four guys now have to drive down to Parsons Avenue hoping to see women on a screen with their shirts off."

He glanced at his computer monitor. I knew he was checking to see if anyone had responded yet to his letters, the ones asking that his family receive that which he believed he had earned. That's one thing that hadn't changed: waiting for good news, hoping. Technology or no, the nervous uncertainty is the same as in the days of Western Union boys and, before that, the Pony Express and smoke signals.

"The first e-mail I ever saw was one that Chuck sent to you," I said.

"How?" Jack said.

"It was in the early 1990s, and I was in town and he and Joyce and I went out to dinner," I said. "You and Janice had moved to Minneapolis. He was trying to explain e-mail to me, and I couldn't fathom it, so when we got back to his house he started up his computer. I'd never seen America Online before."

"And he sent me an e-mail?" Jack said.

"He was drunk," I said. "He typed in 'You suck,' and sent it to you."

Jack laughed. "Sounds right," he said.

"I was very confused," I said. "I kept asking him what would happen if you weren't sitting at your computer to get the note. I didn't understand that it would wait for you, like a letter."

"It's a good thing Chuck never wanted to become Sally Flowers," Jack said.

"What do you mean?" I said.

"'Hello to all our shut-in friends,'" he said. "'You suck.'"

"HELP ME MOVE THOSE THINGS," HE SAID.

The oxygen tanks were askew; they were in this room just for storage, but the way they had been left violated his sense of order. So I walked over to them and he told me how he wanted them arranged next to the wall.

"That's better," he said. "Now no one will trip on them."

I was by a window overlooking his front yard and the street, and he came around from behind his desk to join me.

"You know what this view reminds me of?" he said,

looking out. "What it looked like from the window of your room, when we were kids."

"Well, it's the same part of town," I said.

"Not just that," he said. "You look out there and everything just seems so . . ."

He paused.

". . . self-contained," he said.

I thought I knew what he meant. When we were boys—before we, and the world itself, became so thoroughly connected by, and defined by, the myriad media images that would soon enough surround us—the universe, on many days, seemed no bigger than what we could see from a second-floor window. Part of it was because we didn't know any better—when we first knew each other, we didn't even know how to read—but mostly there must have been something comforting in telling ourselves that what was right there in front of us was all that really mattered.

"When we were born there wasn't even any such thing as television, at least in houses," Jack said. "Sometimes I think about what it must have been like for people not to be looking at the same TV images every night, the same shows, wherever they are in the country. What it was like when you didn't assume that everyone was looking at the same things you were looking at and hearing the same things you were hearing."

"Maybe your reference points then were just your own family and friends," I said. "But probably not. There was network radio before television. And books, and magazines. People had ways they could connect with the outside world."

"Yeah, but they had to work a little harder at it," he said. "I'll bet you it didn't bombard them all the time."

"There was that one little last period of time when we were in our twenties and thirties, right before all of this kicked in completely and overwhelmed everything," I said. "No personal computers. No cell phones. No World Wide Web. No caller ID or call-waiting."

"It was the last days of the busy signal," he said.

"Pretty much," I said. "The last days of not being able to be reached."

He leaned against the windowsill, gazing out, and said:

"You know what you were talking about before? About being able to get anything you need in the world just by calling up a site on the computer and placing an order?"

"It's true, right?" I said.

"Yes," he said. "But I'm standing here looking out at the street, and . . . I don't know why I'm thinking of this, but do you remember when we really wanted those pickles? And your mom got them?"

I hadn't thought about that in years. There was a drive-in restaurant during the carhop days—it was called the Town House, it didn't last for long, it was the Main Street competitor to the Eastmoor Drive-In—and there was something about the pickles on their hamburgers, something so tasty, almost addictive . . .

"Those little round dill pickles," I said.

"Right," he said. "And we talked about how much we loved them, and our moms bought pickles for us over at Kroger's grocery store, pickles that they thought would taste just like the ones at the drive-in. But they didn't. And our moms went back and bought more pickles at Kroger's, and they still didn't taste the same."

"Didn't my mom manage to get the real things?" I said.

"That's what I was just thinking about," he said. "She went over to the Town House, and she walked inside and asked for the manager, and she said she wanted to buy the pickles they used on their burgers. And the guy told her they didn't sell pickles retail—if you wanted to eat their pickles, you had to go there and sit in the restaurant or in your car and order a hamburger or a cheeseburger with pickles."

"But she got them, right?" I said.

"Yep," Jack said. "I don't know how she talked the guy into it, but he sold her this big tin of pickles—you know, an institutional container, not like you see in a grocery—and she brought it back to your house, and . . . man! . . . the way those burgers tasted with those pickles on them."

"Why are you thinking about that now?" I said.

"Because you're right about the computer—if I wanted to, I could probably find any kind of gourmet pickles I wanted on the Internet. I could read about what other customers thought of them, and I could look at color pictures of them close up, and I could find descriptions of how they taste, and if I wanted to pay a little extra I could have them overnighted so that you and I could have them for lunch tomorrow.

"But there was something about the world seeming a little smaller, and it feeling like your mom had won the Olympics by driving down to the Town House and talking that guy into selling her the pickles. Look at me, remembering it now. Do you think I'd be remembering it if she'd placed an order on the Internet?"

"That would have been a trick," I said. "Telling her in 1957 to turn her computer on and go to pickles.com."

"You know what I mean," he said.

He looked out the window some more, as if afraid that a shade would soon be pulled down over the view.

THE PHONE RANG TWICE, THEN STOPPED. We could hear that, downstairs, Janice had answered it.

"He's had a tiring couple of days," we heard her voice say. "Thanks so much. I'll tell him for you. He's resting right now."

"She's been doing that for me," he said. "I just don't want to talk on the phone. She's screening my calls."

"Who would have thought it?" I said.

"What?" he said.

"When your sister Helen first told us we had the girlie hang-ups, who would have ever thought that you'd have someone hanging up for *you*?"

He beamed. "Janice is the girlie? Doing the hanging up?"

"Something like that," I said. "I'd phrase it a little differently, if you should mention it to her."

"She's really taking care of me, isn't she?" he said.

ONCE, WHEN HE WAS IN HIS FORTIES, HE told me about a woman who had come on to him on a business trip.

He'd met her in the course of his job a few times over the years, and on this particular trip they'd gone out to a business dinner with some other people and she had made it clear she was available.

He was tickled about it when he told me; he was surprised when it happened, and it made him feel good that she was so direct in what she wanted, and that what she wanted was him.

He had let her know that it was not an invitation he could take up. (Jack, being Jack, did his best to let her down easily, as if chivalry on his part in such a circumstance was necessary.)

I didn't even have to ask him if he was telling me the whole story. I knew he had stopped that evening in its tracks right where he said he'd stopped it.

"She was really attractive," he'd said to me.

"She probably thought she was propositioning Steven Spielberg," I'd said.

Which was easier for me to say than what I had felt, which was:

Way to go.

You did it right, once again.

NOW, FROM DOWNSTAIRS, JANICE CALLED: "That was . . ."

She said the name of the person who had phoned.

"I said you couldn't talk," she called upstairs.

"I heard," Jack called back to her, although speaking loudly had become difficult for him.

To me, he said: "I don't know what I would have done without her these past months."

"Luckily, you didn't have to find out," I said.

"Luckily is right," he said.

HE SAID HE HAD FORGOTTEN TO DO SOMEthing.

"I don't know why I can't remember things," he said.

But of course he was remembering everything—everything from throughout his life. It was current details—things that were going on right now—that seemed to be giving him trouble.

"I should have written myself a note," he said. "I wanted to make this call first thing this morning."

He went back to the desk and punched a number into the telephone. He didn't ask me to leave, so I was there to hear his end of the conversation.

A charitable organization of which he was a member had held a vote on whether to contribute funds to the families of ten people, mostly immigrants from Mexico, who had been killed in a devastating fire on Columbus's west side. Jack wanted to know how the vote had come out.

"*What?*" I heard him say to the person on the other end of the line.

A second or two passed, and then he said: "That's terrible. That's just terrible."

I found out later what had happened: The organization had already given its yearly donation to an alliance of Hispanic groups that, among its many activities, was helping out with the fire relief; the bylaws of Jack's organization apparently prohibited it from funding more than one grant for an individual cause. So, on that technicality, the board of Jack's organization had turned down the gift to the fire survivors.

"What was the vote?" I heard him ask. "How close was it?"

Then: "I want you to send me the names and phone numbers of everyone on the board, and how they voted."

In the end, he would prevail; he would come up with a way to funnel a new donation through a local bank that was collecting money for the victims' families. All it would take was a little creativity in getting it done.

At the moment, though, he was hearing for the first time that the vote on assistance had been no.

"I don't understand something like this," he said to me after he'd hung up. "These poor families have nowhere to live, and we find some rule that doesn't let us help them."

He walked back over to the window and looked out with me toward the quiet streets.

"Winter's coming," he said.

THE NEXT DAY, BEFORE GOING TO SEE him, I walked through Bexley by myself.

On Drexel Circle, at the corner of Drexel Avenue and Broad Street, I stopped to look at the war memorial.

The Circle is set back off the street. It's more decorative than functional; motorists don't really cut through there. The war memorial, a modest stone obelisk, sits on a patch of grass.

The names of people from the town who have served in America's wars are carved into the memorial's four sides. I wanted to look for my father's name.

It had always bothered him, although he had never complained about it in public, that he had been left off the memorial. He hadn't been born here—he had grown up in Akron, in the northern part of the state. But he had been living here by the time the war began; according to him, he had been the very first Bexley resident drafted into World War II, and had served in the infantry in North Africa and Italy before coming back to begin raising his family.

So being slighted on the war memorial had made him feel kind of bad. Then, toward the end of his life, there had been an article in one of the local weeklies asking families of veterans if they knew of anyone who had served in the military during wartime, and who had inadvertently been left off the memorial. My dad had responded; he had told them that serving in the Army during World War II had constituted his proudest years, and that it would mean a lot to him to be included.

His name went up before he died. Now, on Drexel Circle, I looked for it.

It took me a few minutes, but there it was.

We all, it seems, yearn for something that will let people know we once were here.

I WALKED TO JACK'S AFTER THAT. JANICE said that he was in bed, and that a nurse had come to talk with him.

I said I could come back later, but she said she was sure Jack wouldn't mind if I was in the room.

So I climbed the stairs. He was breathing through the oxygen tubes, and the nurse was explaining to him what most likely lay ahead.

"This is my friend," he said to her as I entered the room, and she and I shook hands.

I sat and listened, and after she had spoken to Jack for a while she asked him if he had any questions.

"I have one," he said. "I weighed myself, and I'm down to a hundred forty pounds. I know that, if I go down to one twenty-five, I probably can't live. So I want to get up to a hundred and forty-five pounds as soon as I can. How should I do that?"

The nurse told him, as delicately as she could, that there might not be a way for him to gain the weight.

"It's not that you're doing anything wrong," she said to him. "But because of the cancer, it just may not be possible for you to put back the weight."

"Just a hundred forty-five," he said to her. "If I can get up to that, I think I can live. Tell me what I can do to try."

One more try—one more nice thing he wanted to do for his family.

Twenty-four

TIME BEGAN TO FEEL COMPRESSED.

ONE AFTERNOON CHUCK CALLED ME AT the hotel where I was staying to say Jack had asked that the three of us go out to dinner. He said Jack wanted to eat somewhere other than in Bexley.

"Do you think it's because he doesn't want people he knows to see him with the oxygen?" I asked.

"I don't think that's it," Chuck said. "He told me he's not bringing the oxygen. I think he just wants to see some different sights."

When Chuck picked me up that night, Jack was already in the car. We drove to a restaurant north of downtown Columbus, somewhere I'd never been.

"This is good," Jack said as we entered. "It's not filled with people who are going to come up and talk."

So that was it; he preferred to have a meal with just the three of us present, and not have to explain to people—people with the best intentions, people who had known him for years—how he was getting along. He was looking more pallid, and I don't think he wanted to put himself on that kind of display.

It was a quiet meal. He kept getting too warm, and then too cold. He would tell Chuck and me about it, and we would ask the waiter to adjust the temperature in the restaurant, or to open a door or close a window. We could hear it when Jack breathed; there was a watery sound in his lungs, and it was audible even across the table.

He was silent a lot. On the way home, Chuck asked if he wanted to stop at the Bexley Monk, and I was very surprised when he said yes.

The Monk came into existence when Bexley finally modified its laws so that restaurants were allowed to serve alcoholic beverages with meals. For all the years of our growing up, the town had been essentially dry; it was no coincidence that the Top had been built just outside the city limits. But when the law had changed, the Bexley Monk had opened, back at the far reaches of the parking lot of a little shopping strip.

It wasn't the most glamorous location, but the food was good and the bar was a popular gathering point just about every night of the week. That's why I hadn't expected Jack to say he wanted to go there. If he didn't wish to deal with people he knew, the Monk was the worst place for him to be.

"We might as well," he said. What he didn't have to say is what Chuck and I both understood: Evenings when he felt strong enough to go out might be coming to an end very soon, and he wanted to sit in the Monk one more time.

The place was crowded, but there were some available stools at the bar, and we took them. It was noisy; I knew we wouldn't be staying long.

"I was supposed to call and get my CAT scan results this afternoon," Jack told us.

He'd had another full-body scan a day or two before; the purpose was to tell him if the new chemotherapy medicines were slowing the progress of the cancer.

"How did you come out?" Chuck asked. He, like I, found it odd that Jack had waited until this late in the evening to mention it to us.

"I didn't call," he said.

"You didn't?" I said. "Why?"

"Because if the answer was bad, I'd be thinking about it all night tonight," he said. "I wanted to have the night with you guys without knowing the answer.

"I just wanted to enjoy the evening."

HE WAS SITTING BETWEEN CHUCK AND me, and when I looked over past Jack the worry in Chuck's face made him appear older than his years.

For just a moment I had a vision of Chuck and me being here—in this place—and Jack being gone. Such a setup had happened remarkably infrequently over the years. Most of my memories of Chuck include Jack in the picture.

There was one ABCDJ night, though, when I know for sure that Jack was absent. I'm not certain why he wasn't with the rest of us—it was a winter evening, maybe he was home with his father—but what I recall, clear as a photograph, is Chuck, Allen, Dan and me walking along snow-fringed sidewalks, wearing Beatles masks.

They had been provided by Chuck's dad, a master at buying up carloads of merchandise at distressed prices, just at the moment the sellers were most eager to unload. Such a moment had come after the Beatles' first flush of saturation-level success in the United States; there were so many Beatles products on the market that first year, consumers were getting a little surfeited, if not bored. So Sol Shenk, at rock-bottom prices, had agreed to purchase an unwanted warehouse lot of plastic masks bearing the likenesses of all four Beatles. He'd bought enough of them that, even by marking them up just a nickel or so, he knew he could turn a profit.

He had given four of the masks to Chuck, and on that winter night the four of us, minus Jack, had decided to go on a date with nature. The concept of "dates with nature" had originated with Dan; it was his term for walking around aimlessly outdoors on nights when any girls you might have asked out had told you no. The customary dialogue: "Hey, Dan, did you have a date tonight?" "Yeah—a date with nature."

On this night we slipped the elastic bands over the tops of our heads and adjusted them in back until the masks were held firmly in place against our faces. The masks, from the inside, had that newly-molded-plastic smell; each of us couldn't see himself, but through the eyeholes we could see the others. Chuck was Paul McCartney, Allen was John Lennon, Dan was Ringo Starr—we had put on our masks beneath the pale illumination of a streetlight amid the falling snow, and I hadn't looked closely at the front of mine, but after seeing the others I knew I had to be George Harrison.

We walked up and down the frigid streets, plastic Beatles on our date with nature, our voices muffled. When I try to visualize it now, I see the grinning likenesses of the four musicians on the fronts of the four masks, and in memory the four of those faces are so achingly youthful. But we were even less far along in life than they; we were pretending to be four men who were irrefutably older than we were. They—the newly famous men on the painted masks—were the seasoned grown-ups; we were the kids.

And Jack wasn't there. That's what I remember most of all, because it was so out of the ordinary. Tonight I looked over at Chuck on his barstool, his face lined and weary; I looked at Jack sitting next to him. We had been so young behind the masks, and now here we were, and I tried not to think about Chuck and me coming back here some night alone.

AT THE BAR, JACK TOLD ME SOMETHING that he'd already told me twice in the last twenty-four hours. Exact same story, for the third time in a day.

He was loopy again—forgetful. It was happening more and more often—the loopiness, as it had been defined by him and Chuck, was getting more pronounced. I could sense that he had no idea he had told me this particular story, in almost the same words, twice before. I didn't interrupt; I could see that Chuck had heard the story, too, but we just listened until Jack was finished.

IN THE MORNING HE AND JANICE CALLED for the CAT scan results.

"Bad," she said to me on the phone.

HE WANTED TO BE ALONE THAT DAY, BUT the next afternoon he told me that he was going to watch the Ohio State football game from bed, and that he'd like me to sit with him.

The nurse was in his bedroom when I got to his house; the television set was tuned to the pregame show.

The nurse—I felt a little sympathetic for her, she was in a room with two guys who had known each other all their lives, she had to have been feeling like the outsider in the group—was doing her best to keep Jack's spirits up, and as part of that effort she said to him:

"Go Bucks!"

He was languid from the effects of the day's medication, and there was a brief delay between when he heard her words and when he looked over at her to acknowledge them.

"Are you going to wear your 'Go Bucks' hat?" she asked in a cheery tone.

My goodness, I thought. It's an Ohio State football Saturday, and he's being talked to as if he's an old man. It wasn't her fault; she wasn't intending to patronize him. The desperately ill man in the bed was the only version of Jack she'd ever seen.

"My hat?" he said to her.

"I thought I saw you had an Ohio State hat when I came in the house," she said.

"Oh," he said. "Downstairs."

The ball was kicked off, and, hearing cheers, we turned toward the noise coming from the box.

IT WASN'T MUCH OF A GAME, AND JACK was doing all right, so after a while the nurse asked us if we'd like to be alone to talk. We didn't want to offend her; it must be very difficult, coming into people's homes, trying to do your job in a setting that's well-known to everyone except you, among people who appreciate your presence but wish fervently that you didn't have to be there. We said that if she wanted to take a break, that would be fine.

So she went to get some coffee or something, and Jack shook his head at me—what a mess, the gesture seemed to indicate—and then he said:

"Do you think Janice will get married again?"

Nothing could have caught me more off guard.

"Why are you saying that?" I asked him.

"It's just something I've been thinking about," he said.

"Does the idea make you jealous?" I said. "Is that why you're thinking about it?"

"No, no," he said. "I'd be happy for her. I want the rest of her life to be happy."

"Jack, don't spend your time thinking about that," I said. "You and she are still here."

"Well, I can't help it," he said.

Someone intercepted a pass, I think. The roar from the television set grew a little louder.

THE NURSE RETURNED. WE ALL WATCHED the game, and then she read to Jack—his ability to concentrate on written words had diminished drastically, but he still hungered for the kind of narrative contained in books and magazines; television left him feeling empty—and when she had finished he returned to the football.

A commercial came on during a time-out, and one of the actors was a man wearing a tuxedo.

"Was it the pinstriped one your dad wouldn't let you have?" he said, smiling.

There it was again: The loopiness might be making him hazy on things that were going on right now, but on something like the pinstriped tuxedo—or the pickles from the Town House Drive-In—he was as keenly tuned as ever.

"I think it was seersucker," I said.

"I don't think there's any such thing as a seersucker tuxedo," he said. "Mine was madras."

"It was just the coat part of the tuxedos, right?" I said. "Not the pants."

"Obviously," he said. "If the pants were madras, it wouldn't be a madras tuxedo. It would be a madras suit."

We'd been invited to some kind of dance, and the invitation had said the event was formal. Formal dances, when we were in school, meant only one thing: O. P. Gallo's.

O. P. Gallo's, with a bustling street-corner store downtown, rented tuxes. Jack and I had gone down to O.P.'s and, feeling potentially natty, had asked the salesman if he had anything available for rental other than the standard black tuxedos.

He did. He showed us what he had. I selected a light-blue-and-white pinstriped model, Jack selected a green-and-red madras. We were to pick them up the next weekend.

"Man, your dad went *crazy,*" Jack said.

The nurse was taking in every word.

"And he hadn't even seen the tuxes," I said. "We didn't even have them yet."

"I know," Jack said. "That was the great part. He was bawling you out just on the principle of ordering a pin-striped tux. He didn't need to see it to hate it."

"He made me cancel the order, right?" I said.

"Oh, yeah," Jack said. "He said he would not pay a cent to rent a pinstriped tux. He said a tux that wasn't all black was not a tux. He said that a tux had to be black and plain because the whole purpose of a tux was to make a man look dignified and elegant."

"Right," I said, "because if our tuxes weren't black, then people on the street in Bexley might not mistake us for Cary Grant and Fred Astaire."

"He wouldn't even let you change your order on the phone," Jack said, laughing. "He made you go back down-town so there would be no possible mix-up."

"Did you have to change your madras tux?" I asked.

"Are you kidding?" he said. "My dad didn't care."

We looked at each other. No words were needed. We'd been having these conversations for fifty years. And I was the one who was going to be left to miss them.

THE LEAVES WERE BRIGHT RED. BIG, brittle red leaves, covering the sidewalks.

They'd turned suddenly, in the weeks just past. They had been green and waxy, secure on their tree branches, and now they'd dropped. They were quite beautiful. I knew I'd seen leaves just like them somewhere.

Were they maple? I picked one up, and as I touched its thin stem I knew where I had found leaves like this before. Right here—on these sidewalks. We had pasted them into scrapbooks. It had been some sort of elementary school project—identify trees by their leaves, collect the leaves and paste them, properly labeled, into scrapbooks for class. I hadn't paid attention to the leaves on the sidewalks in years. But Jack and I, when we were very young, had walked all over town, picking them up from the sidewalks by their stems. I knew it now.

I was on my way to his house again. It was almost time for me to be heading back to Chicago. I followed the route he and I had been taking this year: past the ABCDJ brick, past the hill by the side of the house where he grew up, both the brick and the hill today scattered with those fallen leaves.

It was around noon, and I thought that maybe he would be hungry, so I set off for Rubino's. It's the pizza place on Main Street where our families had been going since the week it opened in 1954. Probably that's why I had never, before this year, set foot inside Pizza Plus, on the site of the old Toddle House: some sort of vestigial loyalty to Rubino's. I was going to order Jack a pizza and carry it to his house; I didn't know if he'd have the appetite for it, but if he didn't maybe Janice or Maren would want it.

The traffic-light boxes made their clanky shifting mechanical hum while the lights, as if copying the maple leaves, switched from green to yellow to red. Maybe traffic lights in every city make that metronomic metallic gears-changing sound, but I'd only noticed it here, and only forever. Maybe it was because here there were few competing sounds. It was a sound as familiar to me as crickets on a central Ohio summer evening, the cricket songs that decorated the air as we'd sit outside and talk late into the night.

Summer was long gone now. Just in the time since I'd been walking today, the temperature had seemed to drop ten or fifteen degrees; it had already been chilly when I started out, but now it was stingingly cold, and a harsh rain had kicked up. My clothes were soaked within a minute.

Such are the hazards of a date with nature. At Rubino's I sat in a back booth as they cooked the pizza; "How's Jack?" they asked, and I said fine, not knowing what else to say. I told them that I was taking the pizza to him for lunch, and this seemed to please them. They asked me to give him their best. By the time I was half a block down Main Street toward the turnoff corner to his house, the white paper that wrapped the pizza was wet and soggy. Probably I should have asked for a box, but we never liked Rubino's in boxes; the white paper was puffed up like a chef's hat, we always thought the pizzas tasted better when they were wrapped that way. Made no logical sense, I know, and by now, what was the difference.

I HADN'T ASKED IN FRONT OF HIM, BUT I had suspected it:

The nurse who had been coming to see him was a hospice nurse. A nurse who is there not to help you get better, but to make your dying as humane and merciful as possible.

Janice told me that these first days of hospice were intended to be a way for Jack to decide how he felt about it; he was free to stop hospice at any point, if he determined that he wanted to go back to aggressive chemotherapy, perhaps even experimental treatments. But as sick as he was, he and Janice, through their doctors, had asked that hospice come out to start him down the path.

The nurse was reading to him again when I arrived from Rubino's. He seemed to be half asleep, the oxygen tubes snaking onto the hospital-type bed in his room.

"You brought me a pizza?" he said, both approvingly—he had always loved Rubino's—and chidingly—did he look like a guy who, at the moment, was ready to dig into a large pepperoni, well done?

"Maybe not the best idea," I said.

"I'm sort of tired," he said.

"I know," I said. "Why don't I come back later. I've been walking around. I can do that some more while you nap."

"Greene," he said, propping himself up on the bed. "What are you wearing?"

I looked down at my clothes. I had on a light windbreaker and a pair of jeans, both wet from the rain.

"What do you mean?" I said. "You see."

"You can't wear that if you're walking around," he said. "It's almost freezing out there. It's raining."

"I'm fine," I said. "I knew it was going to be cold. I've got two shirts on underneath."

"No, no," he said. "I've got a heavy jacket you can wear."

"I don't need it," I said. "This is all right for me."

He called out to downstairs: "Janice?" She didn't respond. He strained to call even louder: "Jan?"

"Jack, I don't need a heavier jacket," I said. "I wear this one all the time."

"You're not leaving without my jacket," he said. He sat up further, pushed the oxygen tubes to the side of his face, and tried to shout for her: "Jan?"

"Don't do that," I said. "It can't be good for your voice to be straining it like that."

"I won't do it if you promise you'll take my jacket," he said.

Janice appeared in the doorway; she had heard his last beckoning.

"I have a heavy black jacket down in the back closet," he said to her. "Look at what Greene's wearing. Don't let him go out without my jacket on."

She looked at me and shrugged. "You heard him," she said.

He lay back down. "Promise me, Greene," he said.

"I won't leave without the jacket," I said.

He drifted off to sleep; Janice and I walked down the stairs. She went to a closet in a hallway off the kitchen, and retrieved a black jacket made of winter-ready material.

"Wear it," she said. "You know he's going to ask me about it when he wakes up."

So I did. The temperature seemed to have dropped another few degrees, the air was still wet and raw, and I left the house as he rested. Full circle, I thought. He's still looking out for me. At the very beginning, when we first met, more than fifty years before: *Bob's hurt.* And now, weak as he was, as he approached the end: *You're not leaving without my jacket.* I don't know how a man becomes so lucky, so blessed; I don't know what a man does to deserve such a friendship.

Twenty-five

I KEPT HAVING THESE THOUGHTS OF ALL the years that had led up to here. I was loading every day and every night with memories of the life he'd led. He told me he was doing the same thing: trying to clear the decks, making sure he did his best to recall everything he could.

"The first time we saw television in color," he said on the phone one night when I was back in Chicago. Breathing was becoming even more difficult; I could hear it across the miles.

"We saw it together," I said.

"I know," he said. "But where was it?"

"My grandmother's," I said. "On Brentwood. She had one of the first color television sets in town. In that sitting room on the second floor of her house."

"Was it *Peter Pan*?" he asked.

"Right," I said. "With Mary Martin."

"I thought so," he said. "That was something. To see it in color. From Broadway, to us."

ONE DAY HE DIDN'T ANSWER HIS PHONE at all until midway through the evening, and when he did I could tell that something disquieting had happened.

"I had a real bad day," he said.

"Did you have to go back to the hospital?" I asked.

"No, it wasn't that," he said. "They came to talk to us about the funeral arrangements."

He had wanted to be involved in the planning. It had seemed like a good idea. But the reality of the meeting, and the talking about the details, had shaken him.

"I thought about all the people who would be sitting there," he said. "And that I wouldn't be there. Of Janice and Maren in the front row, and what was going to be said during the service.

"I could almost see it as we made the plans."

His voice broke.

"So it's all taken care of now?" I said. "You don't have to do anything else?"

"Everything's planned," he said.

And then in the next sentence he said he'd made Chuck and Joyce a music CD to give them for their anniversary.

"A lot of Elvis, a lot of Beatles," he said.

Funeral arrangements and Elvis Presley, all in one day.

"I'll make you a copy if you want one," he said. "But don't tell Chuck yet. It's a surprise."

PEOPLE WANTING TO COME SEE HIM HAD to wait their turn; Janice was doing her best not to wear him out further by subjecting him to endless visits, but so many friends and relatives, some from out of town, were asking to spend time with him that she had to schedule his days.

"I want to see everyone," he told me. "I can't believe how many people want to come over. But I just get so tired."

There was something else about it, too.

"I know why they're coming," he said. "I know why they're in a hurry to get here."

As we spoke every day I thought back to that year when, to keep us from talking, the elementary school teacher had separated us, had moved us apart in her classroom. Now, it seemed, the world was about to separate us, this time for good.

ONE MORNING I CALLED AND HE SAID, "Chuck was here last night."

That wasn't unusual, so I knew he had a reason for bringing it up.

"He came up to my room," Jack said, "and he said to me, 'You know, with all the people who are coming to visit you, we haven't had a chance to talk alone in a while.'"

Jack knew that was true; there had been few times in

recent weeks that people hadn't been coming in and out of the house.

"I told him that if he wanted to talk alone, we could close the door so that it would just be the two of us," Jack said.

Chuck had gone over and closed the bedroom door.

"He sat down," Jack said, "and he told me, 'I think you've done a very good job planning your estate.' He told me he was impressed.

"And then he just started crying."

Jack said he had never seen Chuck cry like that. He said he couldn't stop.

When Chuck finally was able to compose himself, Jack said, he said some very private things meant to allay any fears Jack might be having about Janice's and Maren's future.

What struck me as being so kind—so intuitively right in its tone—was what Chuck had said to begin the conversation.

Chuck was always the one who could do business deals with his eyes closed; Chuck was always the one with the reputation for knowing all the financial angles, all the ways to gain an upper hand in a negotiation. It wasn't lost on Jack that people had a different level of regard for Chuck's business skills than they did for his.

So when Chuck had closed the door and said, "I think you've done a very good job planning your estate," those were the perfect words. Those were words Jack so needed to hear. He didn't know; as he'd struggled with his illness and tried everything he could to get his financial affairs in order, he didn't know how well he was doing at it. Things had become so difficult, and so quickly; he was trying to take care of his family's coming needs at the same time he was trying to stay alive.

And Chuck, by his choice of words, had made Jack feel as if he'd succeeded. By complimenting Jack on how he had planned for his family's well-being, by telling him he was impressed, Chuck had lifted Jack up, had made him believe he had done well. Had made him feel proud.

I wanted to tell Chuck that I knew—that Jack had told me, and that I thought what he had said to Jack was the most considerate and generous combination of words he ever could have chosen.

But because Chuck and I have never been very good at saying anything serious to each other, I knew that, talking to him face-to-face, I'd never be able to find the words. What I wanted to say to him was the same thing he had said to Jack: I wanted to tell him what a good job I thought he had done. Throughout all of this, how magnificent I thought he had been.

But Chuck and I have never been able to look at each other and say such things, so these words, here, will have to do.

"THERE WERE TWO QUARTERS ON THE table near my bed today," Jack said on the phone.

I was becoming accustomed to the sound of his voice mixed with the sound of the oxygen flowing into his nose. There was a halting quality to his cadence.

"I looked at the quarters," he said, "and I thought, that would buy enough gas to get us through a whole night."

On a weekend evening we would pull up to Luke's Shell Station at the corner of Roosevelt and Main, ask for fifty cents' worth of regular, and it was the ticket to whatever lay ahead, it was a passport to everything. Fifty cents had been quite sufficient; fifty cents' worth of gas could take us anywhere we wanted to go by midnight, and get us back home safe and on time. Fifty cents, night after night, was the price of admission for joy.

We'd watch the numbers roll by on the gas pump at Luke's—he'd slow it down as it approached 50, the numbers would speed through the twenties and thirties, and then when it would hit forty-one cents Luke would do something with the pump handle—42, 43, 44, the numbers would move like molasses, and sometimes Luke would let it get to 51 by mistake, but usually he would stop it right on the 50. Not a penny more, not a penny less.

"I saw those two quarters sitting there," Jack said to me. "It made me want to get up and go out cruising the town."

WHERE HE FOUND HIMSELF GOING NEXT would require an expenditure of more than the two quarters on the nightstand.

He wanted to see the ocean. He didn't say *one last time,* but that was understood.

The man in New York with whom he'd done business over the years—the man who had invited Jack and Janice to stay in his Manhattan apartment when Jack was at Sloan-Kettering—also owned a place in Florida. He sent word: If Jack wanted to use it for a few days, that was fine with him. It would be sitting empty otherwise.

Getting down there was going to be a problem. Jack was in bed much of the time now; the oxygen tank was always with him. He could most likely physically get through an airport, and through the lines, and the screening procedures, and the inevitable wintertime delays, but it would undoubtedly take so much out of him that the trip would be worthless. By the time he arrived in Florida he would probably be so wiped out that it would have been better for him to stay in Ohio.

He wanted to see the ocean. He wanted to feel warmth in the air.

Chuck called a company that chartered corporate jets. He asked if they had a small plane on hand with a pressurized cabin—one in which a person breathing with the assistance of an oxygen tank would not feel uncomfortable.

Such a plane was available, for the right fee. Chuck said he would like to rent it.

So on a frigid day in Columbus, at a private airstrip, Jack and Janice and Maren, with Chuck and Joyce, boarded the airplane.

"Oh, Greene, you should see this place," Jack said to me when they arrived. "My bed is next to this open deck. I can hear the ocean as I'm falling asleep."

THEY STAYED THREE DAYS. IT SEEMED TO revive his spirits; he was strong enough to go out for dinner some of the nights, unembarrassed about the oxygen in restaurants where he knew no one other than the people at his table. He told me he wanted to come back to this place by the ocean again; he asked me if I would come next time.

But there was somewhere else he said he wanted to visit first.

"I'd like to go back to Colorado," he said. "To ski."

"Do you think your lungs can take the altitude now?" I asked.

"I think so," he said. "I'll talk to the doctor."

I assumed, when he said he wanted to ski, he meant that he wanted to watch as the others skied.

"No," he said. "I want to ski myself."

Later, when Chuck and I could talk on the phone with just the two of us hearing, I repeated the conversation to him—the part about Jack planning to ski.

There was a short silence, and then, from Chuck: "He really said that?"

"I was surprised," I said. "He said he was feeling better, but I didn't think he was anywhere near that strong."

More silence, and when Chuck spoke again his voice was unsteady.

"Bob, Jack can't walk across a room without having to sit down and rest."

He was dreaming. Jack was awake, but he was dreaming.

I HAD TO BE IN WASHINGTON.

For the last several years, I've had a routine every time I go there.

In the morning I will leave my hotel and walk along the banks of the Potomac River on the District of Columbia side. When I get to the Memorial Bridge I'll walk across it to Virginia, where a roadway leads to the entrance of Arlington National Cemetery.

I'll visit the cemetery, one of the most stirring and beautiful places I've ever been, and I'll just walk the paths and think. It's an overpowering, sanctified place, lovely and heartbreaking at the same time, and I've found few places in my life that are more conducive to quiet and private reflection. The morning walks have become intrinsic to the Washington experience for me.

Usually, before heading back, I'll go to the gravesite of John F. Kennedy, to pause at the eternal flame for a moment, then turn and look at the unobstructed view of the District across the river.

But on this trip, there was somewhere else in the national cemetery where I wanted to stop. I'd asked one of the uniformed guards on duty. He told me where it was. He said to go to the Tomb of the Unknowns; using that as a reference point, he explained exactly how to walk from there to the place I was seeking.

It took me a little while. There are so many grave markers in the cemetery that finding a specific one, even if you've been given directions, can be difficult.

But then there it was. It was at the end of a row of other gravestones, no different from the rest, just as weathered, not any larger or more ornate than its neighbors.

There was a cross engraved at its top, and then the simple, unadorned words.

AUDIE L. MURPHY
TEXAS
MAJOR INFANTRY
WORLD WAR II
JUNE 20, 1924
MAY 28, 1971
MEDAL OF HONOR

I stood for a long time. I was the only visitor.

When I got back to my hotel I called and told him about it.

"It wasn't a big monument?" he said. "They didn't have it set off by itself?"

"No," I said. "It's at the end of a row, next to a road, so I suppose that's on purpose, to make it easy for people to see. But other than that it's no different from any other gravesite there. It's almost as if he didn't want to be considered any more special than anybody else."

"Kind of figures, doesn't it?" Jack said. "Like he didn't want a fuss made over him. Anything else would probably feel wrong."

Twenty-six

THE CALLS BECAME SHORTER IN DURATION.

He was on morphine. He welcomed it, he said. It seemed to let him breathe easier, and it helped him sleep.

He didn't want me to come back until his next round of chemotherapy was over. That's what he was pointing toward—the next chemo. He thought that would put him back on track.

BECAUSE HE HAD ELECTED TO TRY THE further chemotherapy, hospice, for now, was gone. Those were the rules, and the understanding: Hospice was for patients and their families who had come to the informed mutual conclusion that death was the next step. Jack had determined to fight on.

ONE NIGHT, AFTER TALKING WITH HIM for just a minute or so—he sounded so sleepy, so depleted—I asked if Janice could come to the phone.

She picked up, and he put his receiver down and went back to sleep.

"He tells me that he wants to start chemo again next Monday," she said to me, sounding close to tears. "I don't even know how I'm going to get him down to the hospital."

"Do you really think the doctors are going to let him do it?" I asked. "Do you think his body can take any more?"

"I ask him that every day," she said. "And he tells me, 'I can take anything.' "

ON A FRIDAY IN DECEMBER, THE WEEK before Christmas, I was walking beside Lake Michigan in Chicago. It was full winter—the wind was up, the waves were smashing against the rocks. I had been walking for about forty-five minutes, and I was about to turn around to head back.

I pulled my cell phone from my pocket and there was an indication that someone had called. I hadn't heard the ring; between the sounds of the wind, and the water, and the traffic just to the west on Lake Shore Drive, the noise had drowned out the telephone. The number on the cell-phone screen was Chuck's.

It had been just under nine months since the day in Florida when he'd called to tell me Jack was sick—the day when I had been walking next to the Gulf of Mexico. Now, as then, I stood near the water as I called him back.

"Hi," he said.

"You didn't leave a voice mail," I said, "but it says your number called."

"I didn't want to leave a message," he said. "I was going to wait until I got you."

"What's going on?" I said.

"Well, I just got back in town," he said. "I was on a trip. I was at Jack's last night. He isn't doing so well."

"I know," I said. "We talk every day, but it's like he doesn't have the strength."

"It's worse than that," Chuck said. "I know what you mean, because I talked to him every day when I was out of town, and I know how bad the voice is. But when you see him, you'll understand."

"Did something happen?" I asked.

"You'll see when you come in," he said. "Last night there were some people over, and he was downstairs in the living room, and when it was time for him to go to bed we had to carry him up the stairs to the bedroom. He couldn't walk it."

"And there's no hospice anymore?" I said.

"That's another problem," Chuck said. "For two nights now, he's been waking up and telling Janice he has trouble breathing. Last night a few of his friends sat in the room in shifts all night, to be there if he needed help. Janice can't do it all by herself. She's getting no sleep at all."

"It sounds like I should come in right away," I said.

"That's why I'm calling," he said. "I think it's probably getting close. He had a doctor friend who was here today—not the doctor who's treating him, but someone Jack has known for years. I don't know whether we're talking about a week, or weeks, or a month—the doctor friend said there's no way to predict for sure. He said the only time you really can be pretty certain is when you're in the last twenty-four hours, and we're not there yet."

Chuck said that Jack seemed to have entered sort of a holding period. "Maren's in New York visiting some friends for a few days," he said. "She needed to give herself a little break. She's going to come back next week. So I just wanted you to know. It's not like you should come right now, but you might want to think about coming in the next week or so."

"I still think I should come tomorrow," I said. "I'm going to go home and see if I can get plane reservations so close to Christmas. I'll plan to be there in the morning."

"You want me to pick you up at the airport?" Chuck asked.

"No, I'll just take a cab," I said. "Where are you now?"

"I'm outside his house," he said.

"I'll see you in the morning," I said.

AT HOME I STARTED TO CALL AIRLINES and hotels. The voice mail on my home phone gave me a signal that some of the messages I'd saved repeatedly were about to be erased, unless I resaved them.

The way it works is that it plays the old message, and then you have to decide whether to delete it or keep it. So I listened.

The first message was the one from Chuck, back in March, the one I had saved many times since then.

Nice answering machine. This is Chuck. Give me a call. It's about Jack. He's a little ill, and I wanted to explain it to you. So give me a call. Bye.

The next was the one from Jack, the one he'd made on the bus on the way back from the March appointment with Lance Armstrong's doctor in Indiana.

Greene, it's me. Oh, man. Bad news today, buddy. Give me a call when you get a chance. I'll talk to you later.

I saved both calls again, called the airlines, and was able to get a seat on the first flight the next morning.

A FEW MINUTES LATER I WAS PACKING MY clothes and making arrangements to be away, when the phone rang. I could see that it was Chuck calling.

"Hello?" I said.

There was a silence.

"Chuck?" I said.

His voice was unlike I'd ever heard it.

"Jack died," he said.

We both, across the miles, just held the phones to our ears.

"I went over to Main Street for five minutes to pick some things up," he said. "When I got back to his house he was dead."

Janice had been with Jack, Chuck said. Maren, of course, was still in New York.

He had just stopped breathing.

"I can't talk," Chuck said, falling apart.

"I'll be there tomorrow," I said, and hung up, to emptiness.

Twenty-seven

LATE THAT DAY, I CHANGED MY FLIGHT from the next morning to the next afternoon. I knew that Maren would be flying home from New York, and I wanted to give her some private hours with her mother before I arrived.

Chuck called again. "Jack told Janice a while back that he wanted you to speak at the services," he said. "She didn't know if he'd ever said anything to you."

"No," I said. "He never mentioned it. But I figured."

Chuck said that he was going to call Allen in Canton; he asked me if I'd call Dan.

I did. The receptionist at the cold storage plant said he was in the freezer.

I asked that they get him. He came to the phone. We made plans to go to Jack's house together when I got in.

ON SATURDAY NIGHT CARS WERE LINED UP all along Bexley Park Road.

The front door was open a crack. Dan and I saw, as soon as we walked in, that the house was crowded to the walls.

There was a hand-lettered sign taped to the bottom of the staircase, saying that people could take their coats and leave them in the bedroom upstairs.

The last time I was here, he was in that bedroom. It's where he had told me I couldn't leave until I promised to wear his heavy jacket.

Now people were dropping off their overcoats in the same room.

I couldn't make myself go up.

JANICE WAS THERE; MAREN WAS THERE; Chuck was there; Joyce was there. So were dozens of other people whose faces I didn't know.

There was a lot of conversation about finalizing preparations for the next day's services—who would ride with whom, who would sit where, how people would get back afterward.

As everyone moved from room to room, I saw something displayed on a stand next to one wall of the kitchen.

It was a white plate, with writing on its surface. Jack's signature was in the center; the rest of our signatures surrounded his.

Toddle House.

I walked over and touched its edges.

I AWAKENED IN MY AIRPORT HOTEL TO AN ice storm outside the windows.

It was a day like the guardrail day—a day for slippery streets and blowing snow, a day on which no one, if they could avoid it, would want to leave the house.

Dan picked me up and we drove to the services. The parking lot was full; it would have taken a hurricane to keep people from coming to honor Jack, and maybe even that wouldn't have stopped them.

I DON'T KNOW WHAT I SAID WHEN I SPOKE. I hadn't written anything down. I'd only been getting ready for it for fifty years.

The casket was a few feet in front of the lectern. It was closed. I tried not to look at it, but couldn't help myself.

On the way up the aisle at the end of the services, I walked behind it. I wanted to talk to him about it—I wanted to tell him:

Man, Jack, we thought we'd seen everything. But you won't believe this. Do you know where we were today? Take as many guesses as you want—you'll never guess this one. Not in a million years. This one tops them all. Keep guessing. You'll never get it.

Four rows from the back, four rows from the door, sitting on the aisle, was a woman in her seventies.

As Jack passed her, and then as I passed her, I sensed that she was reaching out her hand toward me.

I looked over at her as I walked.

She was Miss Barbara.

I took her hand in mine and she squeezed it, and then we were out the door, and again I wanted to tell him. Guess who was here, Jack. I know you'll get it—think hard. Guess who came to see you today.

I wanted to tell him everything.

EVEN ON A DAY LIKE THIS, A DAY WHEN you wouldn't think it is possible, there was a moment for unexpected laughter.

It came after the services, as everyone was milling around.

Dan was standing there, and he and I were about to go out to his car. People we hadn't seen in years—people with whom we'd gone to school—were coming up to each other and reintroducing themselves, catching up. Two of them walked up to Dan.

"Dan, what have you been doing?" one of them asked him. The standard what's-your-occupation question.

Dan said:

"I'm the safety director of the Millersport Corn Festival."

I actually felt my eyes widen. I could feel my mouth making the word: *What?*

Dan was nodding his head seriously toward the person who had asked him the question.

"You're the safety director of Millersport?" the person said.

"Not *Millersport,*" Dan said, as if stating the beyond-the-obvious. "The Millersport *Corn Festival.*"

I had no idea where this was coming from—whether it was a Reedeep Reeves deal, or what. I had expected, of course, that when Dan had been asked the question, he would say that he was running the cold storage company.

The Millersport Corn Festival? The safety director?

"How did you get elected to that?" the person asked him.

"It's not an *elected* position," Dan said, slowly. "It's an *appointed* position."

"Dan," I said. "Let's get out of here."

We walked over the icy parking lot to his car.

"What was that about?" I said to him. "Were you making it up?"

"No, I wasn't making it up," he said. "I'm the safety director of the Millersport Corn Festival."

"Then why did I never know anything about it?" I asked.

He shot me that grin and that look.

"Greene, I don't tell you *everything,*" he said.

How I wanted to tell Jack. How I wanted to pick up the phone and tell him the story, a new one, Dan being Dan all these years down the line. He would have loved it. It would have lit him up—he would have laughed and the chipped tooth would have shown and he would have shaken his head in delight, the way he'd been doing all his life.

And I knew: There would be moments like this—moments when I wanted to tell him something, something that would make him smile—for the rest of my own life.

AND THAT IS THE MOMENT WHEN I REALized that I probably would—that I probably would be telling him those stories every time something happened that he would love, every time something happened that would make him nod his head, every time something happened that I knew he would want to hear.

This doesn't die—this is the only thing that lasts forever. Friendship is all that is eternal; buildings rise and fall, public men and women come in and out of fame's glare, the years arrive and then silently drift away. But this—this thing that costs nothing, this thing priceless beyond measuring—never ends. No one can take it from you.

The snow and the ice on this day covered the ABCDJ brick, covered Audie Murphy Hill. That's all you could see, if you looked: the new snow, making it appear as if nothing was beneath.

In all the houses on all the streets, it was starting all over again. Friendships forming, some to evaporate, a precious few to endure. They couldn't know it now, the new friends; no one does, when friendship begins. They couldn't know, not at the outset. Miracles don't announce themselves; magic does not arrive with a timetable in its hand.

But it was happening, as it always has, as it always will. This one immortal thing; this one thing you can hold in your heart. The snow blew across Audie Murphy Hill, hiding it from sight, but soon enough the snow and the ice would melt. On a day like this it was hard to keep that in mind, hard to see it, but spring would be here, and then summer, with crickets in the nights. Nights for friends, new and old, to sit outside, letting memories be born.

ABOUT THE AUTHOR

Bob Greene is a *New York Times* bestselling author and an award-winning journalist whose books include *Fraternity: A Journey in Search of Five Presidents; Once Upon a Town: The Miracle of the North Platte Canteen; Duty: A Father, His Son, and the Man Who Won the War; Hang Time: Days and Dreams with Michael Jordan; Be True to Your School;* and, with his sister, D. G. Fulford, *To Our Children's Children: Preserving Family Histories for Generations to Come.*

As a magazine writer he has been lead columnist for *Life* and *Esquire;* as a broadcast journalist he has served as contributing correspondent for *ABC News Nightline.* For thirty-one years he wrote a syndicated newspaper column based in Chicago, first for the *Sun-Times* and later for the *Tribune.* He is a frequent contributor to the *New York Times* Op-Ed page.

Readers may write to him in care of bobgreenebooks@aol.com.